Future of Business in Asia

M. S. Balaji · Jishnu Bhattacharyya ·
Weng Marc Lim · Huey Fen Cheong
Editors

Future of Business in Asia

Exploring Pathways to Tomorrow's Business Landscape

Springer

Editors
M. S. Balaji
RMIT University
Melbourne, VIC, Australia

Jishnu Bhattacharyya
School of Management
Swansea University
Wales, UK

Weng Marc Lim
Sunway Business School
Sunway University
Selangor, Malaysia

Huey Fen Cheong
Universiti Malaya
Kuala Lumpur, Malaysia

ISBN 978-3-031-95939-4 ISBN 978-3-031-95940-0 (eBook)
https://doi.org/10.1007/978-3-031-95940-0

This Springer imprint is published by the registered company Springer Nature Switzerland AG
The registered company address is: Gewerbestrasse 11, 6330 Cham, Switzerland

Foreword

Asia's growing economic influence is reshaping global business research, demanding a shift from Western-focused studies to understanding the region's unique management practices and developmental trajectory. With Asia poised to account for over 50% of global GDP by 2040, its innovative practices and responses to global disruptions such as the rise of AI, climate policies, and geopolitical tensions offer critical lessons for the business researchers and practitioners.

Asian business practices have significantly shaped the global landscape. For example, Japan introduced concepts like consensus decision-making and lifetime employment, while 'tiger economies' such as South Korea and Singapore emphasised education and human development. China's 'Bamboo Capitalism' and India's 'India Way' focused on social missions and employee empowerment, driving economic growth. At the same time, Asia's challenges include demographic shifts and trade impacts, requiring strategic adaptation to sustain progress and redefine industries through global business transformation.

This book explores the evolving business landscape of Asia, organising insights into three inter-related parts.

- Part I: Foundations and Future Perspectives covers economic evolution, leadership innovation, family businesses, and women entrepreneurs in Asia. Highlights include the trajectories of the Asian Tigers and Cubs, the importance of visionary leaders, and the unique facets of family and women-led businesses.
- Part II: Sustainable Business Practices and Finance focuses on sustainability in business and finance, examining Malaysian banks' sustainable finance disclosures and proposing a circular, inclusive business model for Vietnam.
- Part III: Technological Trends and Market Dynamics examines technological and market dynamics, such as extended reality in marketing, digital transformation challenges, supply chain resilience in shipping, and the evolving role of internal auditing in Asia.

The eleven chapters provide detailed research, case studies, and policy recommendations aimed at fostering innovation, resilience, inclusivity, and sustainability within the Asian business ecosystem.

This book accentuates the importance of structured academic inquiry into Asia's evolving role in global business, highlighting contributions from diverse scholars and suggesting policy and managerial recommendations to strengthen the region's trajectory in global business. It concludes by discussing future directions and areas for further exploration in understanding Asian business.

Professor Denis Dennehy (Ph.D.)
Co-Editor-in-Chief of *Communications of the Association for Information Systems*
and Senior Editor of *Information Technology and People*
Professor of Information Systems and Sustainability
School of Management
Swansea University
Swansea, UK

Acknowledgements

The editors thank all of those who contributed to the completion of this book and the *Futures of Asia* series that comprises seven volumes, including this one.

We extend our deepest appreciation to the chapter authors for their hard work and kind cooperation throughout the process. We are also grateful to the individuals and organisations that helped disseminate the calls for papers (CFPs) like the Regional Studies Association (RSA), various academic research groups on social media (particularly Facebook), and our respective institutions, including our faculties and international relations offices. Special thanks go to our assistant, Fatin Najihah, for her invaluable support in the early stages of this project.

We are also deeply grateful for the constructive feedback provided by the peer reviewers of this book, as well as the opportunities, guidance, and support from Springer's commissioning editor, Doris Bleier.

Additionally, we also acknowledge and thank the following scholars for their insightful feedback on individual chapters, both published and unpublished in this volume:

Reviewed Three (3) Manuscripts

1. Ka Shing Charles Ko (City University, Cambodia)

Reviewed Two (2) Manuscripts

1. Ana Mafalda Matias (Polytechnic of Viseu, Portugal)
2. Gabriel Wee Wei En (Swinburne University of Technology, Sarawak Campus, Malaysia)
3. Stephen Homer (Sunway University, Malaysia)

Reviewed One (1) Manuscript

1. Abdul Hayy Haziq Mohamad (University of Technology Sarawak, Malaysia)
2. Anannya Gogoi (Indian Institute of Management Kozhikode, India; Rajagiri Business School, India)
3. Gesualda Iodice (University of Naples Federico II, Italy)
4. Manisankar Datta (Indian Institute of Technology Delhi, India; M. S. Ramaiah University of Applied Sciences, India)
5. Mon Mon Aung (National University of Union of Myanmar, USA)
6. Natasha Banks (St Peters Lutheran College Springfield, Australia)
7. Sachin Kumar (ABV-Indian Institute of Information Technology and Management Gwalior, India)
8. Surbhi Pahuja (Dominion University College, Ghana)
9. Zahid Hussain (Shaheed Benazir Bhutto University, Pakistan)

* *We also thank peer reviewers who opted for anonymity.*

Again, we extend our deepest gratitude for the support, time, and expertise of these individuals. Their contributions were instrumental in publishing this book. In case we have unintentionally overlooked any individual or organisation that contributed to this project, we express our sincere appreciation and gratitude as well.

Editors' Note

The clusters of future studies are based on Fergnani (2019, 2020), who identified six (6) key research/thematic clusters from reviews of future studies since 1968: 'corporate foresight; past and futures; humanity at the limen; environmental futures; post-normality and complexity, and technological trends'. While the chapter topics present different aspects of business in Asia, the clusters of future studies show different dimensions of futures and future studies.

References

Fergnani, A. (2019). Mapping futures studies scholarship from 1968 to present: A bibliometric review of thematic clusters, research trends, and research gaps. *Futures, 105*: 104–23.

Fergnani, A. (2020). Mapping fifty years of futures studies scholarship (1968–2017). In R. Slaughter & A. Hines (Eds.), *The Knowledge Base of Futures Studies 2020* (pp. 24-47). US: Association of Professional Futurists and Foresight International.

Titles in This Book Series

Future of Business in Asia: Exploring Pathways to Tomorrow's Business Landscape
Future of Cities in Asia: Future Direction of Urban Planning, Technology and Environmental Sustainability
Future of Economy in Asia: Geopolitical Diplomacy, Technological Revolutions and Sustainable Pathways
Future of Education in Asia: Sustainability, Technology and Internationalisation
Future of Healthcare in Asia: Advancing Equity, Ethics and Technology in Asian Healthcare
Future of Media in Asia: Artificial Intelligence, Digital Technology and Media Practice
Future of Work in Asia: Transforming Work Models, Workforce Dynamics and Workplace Technology

Contents

About the Editors

M. S. Balaji is an Associate Professor of Marketing at RMIT University, Australia. Dr. Balaji's research interests include sustainability marketing, retail marketing, and services marketing. He has published over 90 peer-reviewed articles in journals such as *Journal of Retailing, Tourism Management, Journal of Travel Research, Journal of Business Research, European Journal of Marketing, International Journal of Hospitality Management, Technological Forecasting and Social Change* among others. He serves on editorial board of several journals including *International Journal of Contemporary Hospitality Management, Journal of Service Theory and Practice, Services Marketing Quarterly, International Journal of Advertising* and *Journal of Research in Interactive Marketing.*

Jishnu Bhattacharyya is a Lecturer (Research) in Marketing at Swansea University's School of Management, United Kingdom. He obtained his Ph.D. in Marketing from the University of Nottingham and previously held the position of Project Scientist at the Indian Institute of Technology Delhi. Dr. Bhattacharyya's research portfolio encompasses a broad range of interconnected domains, including sustainability communication, behaviour change and sustainability, digital marketing, human–robot interaction (service failures), and the psychology of human–AI interactions. His scholarly endeavours are driven by a commitment to integrating rigorous research with pedagogical practice and leveraging the discipline of marketing to foster positive societal change. He is a staunch advocate for data-centric, analytics-driven, and evidence-based approaches to marketing. His research has been published in leading journals such as the *Journal of Travel Research* and the *Journal of Business Research.*

Weng Marc Lim is a Distinguished Professor and the Dean of Sunway Business School at Sunway University, ASU-Cintana Alliance Global Partner Affiliate Faculty at Arizona State University in the United States of America, and Adjunct Professor at Swinburne University of Technology in Australia and Sungkyunkwan University in the Republic of Korea. Professor Lim is Record Holder for 'Youngest Professor' in the Malaysia Book of Records and features in the Academy of Sciences Malaysia's Top Research Scientists Malaysia and JCI's Ten Outstanding Young Malaysians.

With over 100 publications in prestigious 'A*' and 'A' ranked journals, Prof. Lim's research has been cited more than 30,000 times. As a thought leader, Prof. Lim has led high-level discussions at AppliedHE, Association to Advance Collegiate Schools of Business (AACSB), Association of MBAs (AMBA), Business Graduates Association (BGA), Quacquarelli Symonds (QS), Times Higher Education (THE), United Nations Educational, Scientific and Cultural Organization (UNESCO) and Principles of Responsible Management Education (UNPRME), and World Economic Forum (WEF). Professor Lim holds a Bachelor of Business and Commerce with First Class Honors and a Doctor of Philosophy from Monash University, and multiple certificates post-Ph.D., including a Graduate Certificate in Learning and Teaching (Higher Education) with High Distinction from Swinburne University of Technology and a Certificate in School Management and Leadership from Harvard University.

Huey Fen Cheong is a Senior Lecturer at Universiti Malaya, Malaysia. Her academic interests have been multidisciplinary, mainly linguistics, sociology, education, marketing, and psychology. Driven by a cultural and postcolonial perspective, she collaborated with a group of established scholars to initiate this *Futures of Asia* series (which contains this book) and the earlier *COVID-19 in Malaysia* series. These initiatives aim to promote Global South research and publications, especially on significant events such as the COVID-19 pandemic, post-pandemic normality and transformations, and the rise of AI, which equally matter to non-Western regions and deserve equal scholarly attention (and acceptance) beyond Western-centric frameworks in academia.

List of Figures

List of Tables

Chapter 1
Introduction: Future of Business in Asia

M. S. Balaji and Jishnu Bhattacharyya

Abstract This chapter highlights the growing influence of Asia in business. Driven by economic powerhouses such as China, India, and Japan alongside Vietnam, Thailand, Indonesia, and Bangladesh, Asian countries contribute to over 40% of the global GDP and are projected to increase to 60% by 2030. Despite its prominence, business research has predominantly focused on Western markets. This introduction chapter provides a foundation for the book—Future of Business in Asia—by highlighting the business landscape in Asia. It outlines the key themes explored in this book and sets the stage for how businesses, policymakers, and researchers can analyse the business practices in Asia and leverage the insights for strategic growth.

Keywords Asia · Asian business · Emerging markets · China · India · Future perspectives · Economic powerhouse

1.1 Introduction

Despite a rapidly evolving global economy, shifting power dynamics, and the rise of new middle powers, business research remains focused on mature Western markets in North America and Europe, which have historically dominated the world economy (Bruton & Lau, 2008; Khilji, 2012). This narrow focus risks overlooking Asia, the world's largest and most populous continent, home to 60% of the global population and economic powerhouses like China, India, and Japan, alongside emerging economies like Vietnam, Indonesia, and Bangladesh, which are paving the way for the region's future dominance in global business (Boudreau et al., 2025; Tyagi, 2024). As Asia is set to account for over 50% of global GDP by 2040 (Tonby et al., 2019), its economic influence and distinct institutional contexts demand greater attention in

M. S. Balaji (✉)
RMIT School of Economics, Finance, and Marketing, RMIT University, Melbourne, Australia
e-mail: balaji.makam@rmit.edu.au

J. Bhattacharyya
School of Management, Swansea University, Swansea, UK
e-mail: jishnu.bhattacharyya@swansea.ac.uk

M. S. Balaji et al. (eds.), *Future of Business in Asia*,
https://doi.org/10.1007/978-3-031-95940-0_1

business research. Its unique management practices, shaped by national culture and contextual factors, have the potential to challenge and redefine existing paradigms (Lau, 2007; Leung, 2007). Moreover, the global shift in economic power is replacing Western dominance with increasing recognition of Asia's central role. The once-theoretical '*Asian Century*' is now a reality that business research can no longer afford to ignore (Mahbubani, 2022; The Economist, 2021). Understanding this shift is essential for grasping Asia's growing influence and recognising the lessons it offers the world.

Asia's significance in business research extends beyond benefiting the region itself—the world has much to learn from it. Understanding Asia is essential for global business growth and expansion (Shankar, 2024), especially as decades of progress have placed the region at a pivotal juncture (Aggarwal & Bharadwaj, 2024). This urgency is heightened in an era of disruptions—spanning natural and humanitarian crises, breakthroughs in generative AI, transformations in healthcare systems and economic structures, evolving climate policies, and the complex China-U.S. dynamic (Lewin & Witt, 2022). These shifts raise key questions: How will they shape the future of Asian business, and what can the world learn from Asia's response? The thriving scholarship also highlights the region's deep reservoir of knowledge, historical experience, and innovative practices with significant implications for global business and management (Liu, 2024). Asia's rapid industrial transformation is already reshaping industries—China's DeepSeek, for instance, is challenging Western dominance in AI, signaling a potential shift in technological leadership (Cieslak, 2025). This is just one of many disruptions emerging from the region. Moreover, Asia's developmental trajectory has provided valuable lessons in crisis management and sustainable development (Liu & Froese, 2020). Given these dynamics, as Asia navigates a complex global landscape, its experiences offer critical lessons for seizing future opportunities.

> Asia will be the furnace in which a new era is forged—and may experience heightened versions of global challenges. (Seong et al., 2023)

Understanding and analysing these shifts requires structured academic inquiry and informed discourse. In this book, we aim to foster rigorous academic discourse on Asia's evolving role in global business by tapping into ongoing developments. Our contributors present diverse perspectives on why the region's future holds promise and offer policy and managerial recommendations to further strengthen its trajectory. To close the gap in understanding the future of business in Asia, we begin with a brief background on the region and the definition of Asian business, followed by an overview of the key findings from our panel of authors, who examine various aspects of Asian business, grounded in solid literature. While we cover major contemporary topics, we acknowledge that no single book can comprehensively address all dimensions of this vast and dynamic field. Finally, we discuss future directions for Asian business by reflecting on the contributions and limitations of this volume, offering insights into areas for further exploration.

1.2 What is Asia?

The definition of Asia is not universally agreed upon in research (Bruton & Lau, 2008). While some define it narrowly, others take a broader perspective. Geographically, Asia is considered the region east of Europe and Africa, a classification that ensures findings are not influenced by a single culture. However, this broad definition includes the Middle East (e.g., Iran, Israel, Jordan, and Saudi Arabia) and the Caucasus (e.g., Armenia, Georgia, and Azerbaijan), regions typically recognised as culturally distinct from East and South Asia (Bruton & Lau, 2008). Given these distinctions, we adopt a broad but cautious approach, excluding the Middle Eastern and Caucasus nations. Additionally, although Russia and Turkey have territories in both Europe and Asia, their economic centres are predominantly in Europe. Research on these nations rarely differentiates between their European and Asian regions, which could introduce confounding effects. To maintain clarity and focus, we also exclude Russia and Turkey from our book.

Asia, as defined in this book, encompasses over 30 nations with a combined population of approximately 3.5 billion (Bruton & Lau, 2008). Its economic influence is expanding rapidly, with Japan among the world's four largest economies and China and India projected to surpass the U.S. in GDP by 2050 (AP, 2024; PwC, 2017; Wilson et al., 2003). India is expected to rank second in global GDP at purchasing power parity (PPP), behind only China but ahead of the U.S. (PwC, 2017). By 2050, six of the seven largest economies will likely be emerging markets, led by China (1st), India (2nd), and Indonesia (4th) (PwC, 2017). Many of the region's fastest-growing economies—including India, Bangladesh, Cambodia, Myanmar, Laos, Vietnam, the Philippines, China, and Mongolia—have recorded annual growth rates between 6.2% and 7.4% (Shankar, 2024). This rapid expansion positions Asia as a prime destination for global businesses, potentially offering greater opportunities for growth than the European Union (Shankar, 2024). By 2030, Asia is expected to account for two-thirds of the global middle class, and 18 of the world's 20 fastest-growing business corridors are already located in the region (Kumra & Yamada, 2025). Asia now contributes 55% of global GDP, surpassing the combined 33% share of Europe and North America, marking a significant economic shift (World Economics, 2024). Business leaders recognise this transformation, with a recent Asian Business Council survey showing that over 80% are optimistic about the region's future (Kumra & Yamada, 2025). The growing presence of Asian firms further underscores this shift—over the past two decades, their representation in the *Fortune Global 500* has increased from 19 to 30%, with 210 of the world's 500 largest companies now headquartered in Asia (Tonby et al., 2019).

> Factory Asia—these two simple words help define Asia's phenomenal economic growth over the past two decades. (Subramaniam & Ng, 2014)

Asia is a diverse and dynamic region with rich cultural, aspirational, and institutional contexts that significantly shape global business today (Naafs & Skelton, 2020; Warner, 2012). While referring to Asia as a singular entity may seem simplistic, the region's immense diversity—home to approximately 2300 languages compared to

Europe's 300—reflects vast cultural and political differences. Economic disparities are equally stark; for instance, Singapore's per capita GDP is 60 times that of Nepal. Despite these differences, Asia functions as an interconnected ecosystem, driven by mutual trade interests and economic interdependencies, positioning the region at the forefront of emerging global opportunities (Seong et al., 2023). The growth story of Asia is no longer defined solely by the traditional economic powerhouses of Japan, South Korea, Singapore, and China. New players like India, Vietnam, and Indonesia are rapidly integrating into regional and global value chains, further strengthening Asia's role in the global economy. As the progress of past decades converges, the region stands at a pivotal moment, poised to shape the next chapter of its economic evolution (Aggarwal & Bharadwaj, 2024; The Economist, 2019). It is, therefore, increasingly critical for businesses to shift their focus to markets that will dominate the world stage in the future (Shankar, 2024).

1.3 The Fascinating World of Asian Businesses

The fascinating and evolving landscape of Asian businesses has long intrigued the West, with Japan leading the way through its unique practices, such as decision-making by consensus, TQM, JIT, continuous training, lifetime employment, and the godfather system (Drucker, 1971; Yang, 1984). This initial interest spread to other East Asian nations like South Korea, Singapore, Malaysia, and Taiwan, often categorised as "tiger economies", where a focus on education and human development played a key role in their economic success (Mok, 2008). In more recent years, China's remarkable economic growth has spurred discussions about its entrepreneurial culture, introducing terms like "Capitalism with Chinese characteristics" (Yergin & Stanislaw, 2002) and "Bamboo Capitalism" (Lenzner, 2011). Similarly, India's business model, known as the "India Way," highlights context-driven practices in large organisations, where the focus shifts from shareholder value to social missions, heavy investments in employees, and empowering decision-making (Bhattacharyya et al., 2021; Cappelli, 2010; Cappelli et al., 2010). This approach has contributed to significant business growth. Notably, Indian companies thrive because of their commitment to social missions and employee investment, rather than despite them (Cappelli et al., 2010). As Asian business practices continue to shape the global landscape, there is an increasing need to study and learn from these practices. This exploration is vital for advancing the "East meets West" dialogue (Chen & Miller, 2011, p. 201) and recognising Asia's significant influence, both as one of the oldest cultures and the largest population base.

We can now confidently say that Asia stands at the forefront of global business transformation, offering immense opportunities for leaders who can overcome challenges and shape the region's future (Kumra & Yamada, 2025). One critical challenge, for instance, is demographic shifts—two-thirds of the world's population now lives in countries with fertility rates below the replacement level, posing significant economic and policy concerns. Japan is already grappling with labour shortages and rising healthcare costs, while China and South Korea face similar pressures, requiring innovative strategies to sustain productivity amid a shrinking workforce (Kumra & Yamada, 2025). Among others, another major challenge is absorbing the impact of the ongoing trade war between the U.S. and China on Asian economies (Rahman et al., 2023). Those who succeed will not only lead their industries but also redefine them (Kumra & Yamada, 2025). While Asia is driving global economic shifts, it also faces intensified challenges, necessitating strategic adaptation (Seong et al., 2023).

1.4 Structure and Overview of the Book: The Future of Business in Asia

Optimism about Asia's future is tempered by the urgent need for change, as leaders recognise that the coming decades will be markedly different from the past 30 years and are recalibrating their strategies accordingly. This book examines these evolving dynamics, building on existing literature and outlining future directions for business in Asia. It aims to expand theoretical understanding and provide a foundation for further research. Furthermore, this book offers unique insights into the rapidly transforming business landscape of Asia. Given the unique challenges and opportunities presented by the growing Asia, this book on Future of Business in Asia is both timely and relevant to understand how the Asian Businesses are navigating these trends. Reflecting on the discussions thus far, we now turn to key aspects of Asia's business future, summarised in the following paragraphs as an overview of the volume's chapter.

This book explores the future of business in Asia by organising the research and insights into three parts following Fergnani's (2019) mapping future studies: Part I: Foundations and Future Perspectives, Part II: Sustainable Business Practices and Finance, and Part III: Technological Trends and Market Dynamics. Each part addresses critical aspects shaping the business landscape in Asia, providing a comprehensive understanding of emerging trends, challenges, and opportunities.

1.4.1 Part I: Foundations and Future Perspectives

This part offers a comprehensive analysis on the present and future perspectives shaping the business landscape in Asia. This part includes four chapters that explore key themes such as economic evolution of Asian businesses, leadership and innovation, family businesses, and women entrepreneurs.

Chapter 2 by Suplico-Jeong et al. (2025) critically explores the economic evolution of the Asian Tigers (Hong Kong, South Korea, Taiwan and Singapore) and the Cubs (Indonesia, Malaysia, Philippines, Thailand, and Vietnam). This chapter utilises a Systematic Literature Review (SLR) to examine the development trajectories of

Asia Tigers and Cubs, corporate foresight in business strategy, sustainability impacts within Asian economic contexts, and policy-business interdependencies in Asia.

Utilising deductive and inductive coding of the 40 articles published between 2010 and 2024, the authors identified nine key themes aligning with Asia's economic trajectory. These nine themes—national development strategies, corporate foresight and strategic planning, sustainability and environmental impacts, technological advancements and digital transformation, policy-business interdependencies, synergies in state-business relationships, corporate foresight as a strategic tool, sustainability business models in response to environmental challenges, and technology and digital transformation as a catalyst for growth—provide a comprehensive understanding of the strategic, economic, and sustainability forces driving the future of Asian businesses. This chapter offers significant implications for policymakers, business leaders, and researchers. Specifically, by enhancing public–private collaborations, investing in corporate foresight and scenario planning, implementing sustainability-oriented business models, prioritising digital transformation for economic resilience, promoting flexible and adaptive policies, and expanding research on transferable insights for global applications, this chapter provides a blueprint for developing Asian economies to manage economic transitions and enhancing global competitiveness.

Chapter 3 by Devasia and Haris (2025) explores the crucial role of leadership in driving innovation in the Asian business landscape. Leadership plays a key role in the business landscape by driving innovation and fostering adaptability in the dynamic market. Effective leaders not only influence strategic direction, decision-making, and organisational culture, but also ensures businesses remain competitive, resilient, and sustainable. This chapter focuses on facets of leadership that ignite creativity and innovation in the Asian business landscape. In doing so, this chapter proposes that leadership catalysts or visionary leaders who go beyond conventional managerial responsibilities to propel organisational innovation through curiosity and lifelong learning, strategic thinking, adaptive leadership style, empowerment and trust, and ethical leadership and resilience.

Through a collection of case studies, this chapter explores how visionary leaders such as Masayoshi Son (Chairman and CEO of SoftBank, Japan), Pony Ma (Chairperson and CEO of Tencent, China), Dhirubhai Ambani (late Chairman of Reliance, India), and Wan Zulkiflee (former President and CEO of Petronas, Malaysia) and businesses such as Samsung Electronics (consumer electronics and semiconductors, South Korea), Infosys (Information Technology and Consulting Services, India), BYD Company Limited (Electric Vehicles and Renewable Energy, China), and Toyota Motor Corporation (Hybrid and Electric Vehicles, Japan) illustrate how Asian businesses and leaders overcome challenges to innovation and navigate uncertainty and foster adaptability to steer their businesses towards sustained competitive advantage. Understanding these leadership strategies outlined in this chapter is essential for shaping the future of business in Asia. It also serves as a roadmap for businesses, policymakers, and researchers aiming to cultivate innovation, build resilience, and drive sustainable growth in an increasingly dynamic and competitive market environment.

Chapter 4 by An (2025) provides an overview of scholarship of family businesses in Asia. As family businesses forms the backbone of the Asian economy and play a critical role in its driving economic growth, this chapter explores the governance, strategy, and succession of family businesses using the bibliometric analysis. Specifically, the Theory-Context-Characteristics-Methodology (TCCM) framework was used to summarise the existing family business landscape in Asia and propose future research directions. The bibliometric analysis of 198 articles published between 1994 and 2023 revealed a growing interest in research on family business in Asia. It recognises that while much of literature on family business has been in the western countries, the unique economic, cultural, and governance aspects in Asia requires further exploration.

The application of the TCCM (Theory-Context-Characteristics-Methodology) framework reveals that previous literature employed micro- (agency theory), meso- (resource-based view and social capital), and macro- (institution) theories to understand family business in Asia. The key areas of family business examined in the literature are succession, regional development, innovation, and market development. Furthermore, family business characteristics such as leadership, ownership, business performance, and social responsibilities are investigated using qualitative, quantitative, and cross-sectional approaches. The TCCM framework is used to identify eight key propositions, namely the need for diversity, equity, and inclusion theories, consumer-facing theories, and process-based approaches to advance the understanding of family business, further exploration of the foundations and dark side of family businesses, analysing the performance of family business beyond firm-level to understand its impact on the eco-system, and the need for application of advanced research design and approaches such as use of unstructured data and machine learning as well as longitudinal design for deeper and valuable insights. These insights of this chapter provide a comprehensive understanding of the current state of literature on how family businesses grow, operate, and influence the business landscape.

Chapter 5 by Gunasegaran (2025) examines the historical milestones and challenges that have shaped women participation in Malaysia business. Following which, it proposes strategies to fosters a more inclusive business environment. While women are underrepresented in key leadership and decision-making roles in Global and Asian businesses, their participation is increasing over the past few decades. Their participation in business brings diverse perspectives, creative ideas, and broader talent pool, leading to improve business performance and increased profitability. Furthermore, greater women participation in decision-making roles has been shown to improve overall performance, making their participation essential for economic growth and societal progress.

This chapter (Chap. 5) traces key policies and initiatives that promoted the participation of women in Malaysia started with the implementation of National Women's Policy (NWP) in 1989, followed by the Multimedia Super Corridor (MSC) initiative in 1996 that sought to enhance women participation in the digital economy. This was followed by the 2011 policy that mandates 30% representation of women in the leadership roles to enhance equity and representation. Other initiatives such as the Mini Estate System (MES) and Cluster System (CS) were undertaken to improve the income-generating opportunities for women. The declaration of 2018 as Women Empowerment year provide further impetus for improving gender diversity and strategic participation of women in business. This chapter utilises a SLR approach to analyse the impact of these policies and initiatives on women's participation in Malaysia business landscape. It argues that while women participation has improved, deeply rooted social norms continue to hinder their participation and growth. The findings emphasise that future increase in women participation depends on digital transformation, policy reforms, education, access to networks and support systems, and development and implementation of new policies and initiatives by the government. These insights will help policymakers, businesses, and researchers to create a more inclusive business environment that drives greater women participation and long-term economic growth and competitiveness.

1.4.2 Part II: Sustainable Business Practices and Finance

This part explores the evolving landscape of sustainability businesses practices and finance. The two chapters included in this part offer a comprehensive understanding of how businesses and financial institutions drive economic growth and long-term sustainability.

Chapter 6 by Joseph et al. (2025) examines the extent of sustainable finance initiatives disclosures among nine selected Malaysian banks and financial institutions. Sustainable finance plays a critical role for ensuring strong economic growth and financial stability by allocating funds for projects that have a positive impact on the environment and society. By supporting projects that prioritise environment and societal impact, sustainable finance helps in reducing poverty, prioritise resource efficiency, and developing a healthy ecosystem. This chapter provides a comprehensive prospective of sustainable finance in Malaysia by analysing the sustainability and integrated reports of nine Malaysian banks and financial institutions for 2022.

Through content analysis, it identifies disclosure practices of these banks and financial institutions across key categories such as environment management, social, governance, principles of reporting, sustainable finance, and sustainability initiatives. The findings of this chapter reveals that while disparities remain across the majority of banks and financial institutions in Malaysia, they demonstrate a strong commitment to sustainability financing initiatives such as green financing, financial inclusion, climate risk management, and adherence to frameworks such as the Task Force on Climate-related Financial Disclosures (TCFD) and the UN Sustainable Development Goals (SDGs). Furthermore, this chapter highlights the need for greater transparency, standard reporting guidelines, and stronger governance to enhance sustainable finance practices in Malaysia and Asia. The insights obtained from this analysis provides key implications for policymakers, stakeholders, and researchers in enhancing sustainable finance initiatives and disclosure practices.

Chapter 7 by Tran et al. (2025) develops a sustainable business model that integrates circular and inclusive practices. The increasing global challenges and need to align United Nations Sustainable Development Goals (SDGs) requires businesses to adopt sustainable practices that address environmental, social, and economic factors. This chapter focuses on Vietnam, a rapidly developing nation, to propose a sustainable business model with the core values of circular and inclusive models. Integrating circular and inclusive values enable businesses to address pressing environmental and societal challenges. By integrating inclusiveness and circular values, the sustainable business model can better align with the expectations of its stakeholders and ensure long-term sustainability and competitiveness. The insights from this chapter provide valuable insights for stakeholders and businesses in developing a sustainable business model as well as transitioning towards a responsible and resilient business model.

1.4.3 Part III: Technological Trends and Market Dynamics

This part focuses on the technological trends and market dynamics, examining how emerging technologies, digital transformation, and strategic governance impact business landscape in Asia. This part includes four chapters that explore emerging technologies in marketing, digital transformation, supply chain resilience and internal audit.

Chapter 8 by Pandey, Sinha, and Kumar (2025) explores the evolving role of extended reality in marketing. With the business landscape rapidly evolving, the integration of emerging technologies like extended reality in business operations and marketing is a natural progression that allows businesses to adapt to changing marketing dynamics and gain competitive advantage. This chapter utilises SLR of 17 articles on extended reality in marketing to provide a comprehensive understanding of the adoption, barriers to adoption, advantages, disadvantages, and theories used in the literature on extended reality in marketing. Through case studies of Lenskart and Ikea in India, this chapter illustrates the adoption of extended reality and their impact on customer engagement and behaviors. It identifies key future research directions in the areas of consumer behavior and psychology, marketing technology and innovation, and market accessibility. These insights from this chapter has important implications for businesses and marketers in developing strategies to efficiently implement extended reality to improve their marketing efforts and increase sales and profits.

Chapter 9 by Chiam (2025) explores the roles of technology and digital transformation in shaping the business landscape in Asia. In the twenty-first century, recognised by its volatility, uncertainty, complexity, and ambiguity (VUCA), technology and digital transformation has emerged as a disruptor and enabler of innovation, efficiency, and sustainability. With businesses in Asia leveraging technology to adapt to changing market dynamics and enhance customer engagement, this chapter the key challenges associated with technological innovation in Asian countries such as technological division, identify theft and online scams, shift of labor force patterns, digital addiction, spread of misinformation, exposure of private information, and cyberbullying. It also identifies contingencies to address these challenges: equitable access to digital technologies, comprehensive regulations, training of workforce, government policies for upskilling workers, promoting healthy lifestyle, and encouraging sharing suitable information. With digital transformation standing at the forefront of the modern business landscape, the insights from this chapter provide a strategic roadmap for policymakers, businesses, and researchers to navigate its challenges.

Chapter 10 by Wu (2025) examines the growing significance of supply chain resiliency and operations in the shipping industry in Asia, specifically focusing on China and Singapore. As globalisation and trade expand, the demand for efficient shipping operations has increased. This has positioned Asia as a crucial player in shipping and logistics. Through a qualitative case study approach, the chapter explores how China and Singapore have developed strategies to strengthen their infrastructure and supply chain resilience of their shipping industry. This chapter reports that China with its network of major and minor ports has adopted advanced technologies, automation, and smart solutions to enhance efficiency and mitigate disruptions caused by unforeseen events such as the COVID-19 pandemic and natural disasters. Desoute its limited geographic size, Singapore has leveraged its strategic location, digital innovations, and global partnerships to strengthen its role as a key maritime hub. This chapter also highlights key challenges faced by both countries, including geopolitical tensions, cybersecurity risks, and climate-related disruptions, while also emphasising the importance of diplomacy and international cooperation in ensuring the resilience and stability of the supply chain. The insights from this chapter help businesses and policymakers adopt similar strategies to ensure sustainability and resilience in shipping and maritime and become competitive in an increasingly interconnected global economy.

Chapter 11 by Suhaily et al. (2025) explores the changing landscape of internal audit in Asian businesses. With digital advancements, globalisation, and regulations transforming the business landscape today, internal audit serves as a strategic function within organisations. It ensures compliance with policies, assess risks, and enhance efficiency. It is the crucial mechanism for ensuring internal controls and governance. This chapter utilises the Assure-Build-Consult (ABC) model to provide a systematic understanding of internal auditing and how it contributes to building robust governance structures, risk management, and ESG strategies. Through case studies in the oil and gas, financial, and retail sectors, this chapter illustrates how the ABC model enhances internal audit effectiveness. By adopting the ABC model for internal auditing, businesses can future-proof their operations, ensuring resilience, sustainability, and long-term strategic success.

1.5 Conclusion

In concluding this introduction chapter, we anticipate a growing body of research on Asian business as the region continues to reshape global markets. This book aims to inform mainstream research and practice by highlighting key challenges and opportunities, ranging from topics such as economic evolution, leadership as a driver of innovation, the significance of family businesses, and the growing role of women entrepreneurs, to sustainable business practices and finance. By expanding theoretical understanding and offering practical insights, we hope to inspire further inquiry into Asia's evolving business landscape.

References

An, J. (2025). Research propositions for family business in Asia: A systematic literature review. In M. S. Balaji, J. Bhattacharyya, H. F. Cheong, & W. M. Lim (Eds.), *Future of business in Asia*. Springer.

Aggarwal, N., & Bharadwaj, A. (2024, June 24). Connecting the dots behind current and future Asian growth. World Economic Forum. Retrieved March 1, 2025, from https://www.weforum.org/stories/2024/06/why-asia-s-time-is-now-whats-fueling-asian-growth-and-what-does-it-mean-for-the-rest-of-the-world/

AP. (2024, February 15). Japan slips to the world's fourth-largest economy, behind the US, China and now Germany. The Hindu. Retrieved March 2, 2025, from https://www.thehindu.com/news/international/japan-slips-to-the-worlds-fourth-largest-economy-behind-the-us-china-and-now-germany/article67847937.ece

Bhattacharyya, J., Dash, M. K., Kundu, S., Sakshi, S., Bhattacharyya, K., & Kakkar, K. B. (2021). No virus on me: The Indian ways of managing the COVID-19 pandemic, marine to mountain. *Asian Journal of Management Cases, 20*(2), 128–156.

Boudreau, D., McDaniel, M., Sprout, E., & Turgeon, A. (2025). Asia: Physical Geography. Retrieved March 2, 2025, from https://education.nationalgeographic.org/resource/asia

Bruton, G. D., & Lau, C. (2008). Asian management research: Status today and future outlook. *Journal of Management Studies, 45*(3), 636–659. https://doi.org/10.1111/j.1467-6486.2007.00758.x

Cappelli, P. (2010). The India way: How India's top business leaders are revolutionizing management. Harvard Business Press. Retrieved February 25, 2025, from https://books.google.com/books?hl=en&lr=&id=KrmGsUh8IK4C&oi=fnd&pg=PA1&dq=Cappelli+et+al.,+2010&ots=cT-p6wqSgH&sig=hAzNBKHiwgu-tXQ1HL-2r3xJa2Q

Cappelli, P., Singh, H., Singh, J., & Useem, M. (2010). The India way: Lessons for the U.S. *Academy of Management Perspectives, 24*(2), 6–24. https://doi.org/10.5465/amp.24.2.6

Chiam C. C. (2025). Digital transformation in empowering innovation and sustainability in the 21st century business environment in Asia. In M. S. Balaji, J. Bhattacharyya, H. F. Cheong, & W. M. Lim (Eds.), *Future of business in Asia*. Springer.

Chen, M.-J., & Miller, D. (2011). The relational perspective as a business mindset: Managerial implications for East and West. *Academy of Management Perspectives, 25*(3), 6–18. https://doi.org/10.5465/amp.25.3.zol6

Cieslak, M. (2025, January 28). DeepSeek shows AI's centre of power could shift away from US. BBC News. Retrieved February 27, 2025, from https://www.bbc.com/news/articles/c9w5d9new0yo

Devasia, S. N., & Haris, S. (2025). Leadership catalyst. Igniting innovation in the Asian business landscape. In M. S. Balaji, J. Bhattacharyya, H. F. Cheong, & W. M. Lim (Eds.), *Future of business in Asia*. Springer.

Drucker, P. F. (1971). What We Can Learn from Japanese Management. Retrieved February 27, 2025, from https://hbr.org/1971/03/what-we-can-learn-from-japanese-management

Fergnani, A. (2019). Mapping futures studies scholarship from 1968 to present: A bibliometric review of thematic clusters research trends and research gaps. *Futures, 105*, 104–123. https://doi.org/10.1016/j.futures.2018.09.007

Gunasegaran, H. (2025). Women in business: A historical perspective and future outlook in Malaysia. In M. S. Balaji, J. Bhattacharyya, H. F. Cheong, & W. M. Lim (Eds.), *Future of business in Asia.* Springer.

Joseph, C., Norizan, S., Rahmat, M., & Enggong, T. S. (2025). Sustainable finance initiatives disclosures by selected banks in Malaysia. In M. S. Balaji, J. Bhattacharyya, H. F. Cheong, & W. M. Lim (Eds.), *Future of business in Asia.* Springer.

Khilji, S. E. (2012). Editor's perspective: Does South Asia matter? Rethinking South Asia as relevant in international business research. *South Asian Journal of Global Business Research, 1*(1), 8–21.

Kumra, G., & Yamada, Y. (2025). Asia's emerging business corridors: New highways to growth, McKinsey. Retrieved March 7, 2025, from https://www.mckinsey.com/featured-insights/future-of-asia/asias-emerging-business-corridors-new-highways-to-growth

Lau, C.-M. (2007). The first decade of the Asia academy of management. *Asia Pacific Journal of Management, 24*, 401–410.

Lenzner, R. (2011). Bamboo capitalism in China the secret of success. Retrieved February 26, 2025, from https://www.forbes.com/sites/robertlenzner/2011/10/09/bamboo-capitalism-in-china-the-secret-of-success/

Leung, K. (2007). The Glory and Tyranny of citation impact: An East Asian perspective. *Academy of Management Journal, 50*(3), 510–513. https://doi.org/10.5465/amj.2007.25525592

Lewin, A. Y., & Witt, M. A. (2022). China's Belt and Road Initiative and international business: The overlooked centrality of politics. *Journal of International Business Policy, 5*(2), 266–275.

Liu, Y. (2024). Reviving the aspiration, fostering impactful research, and contributing to the sustainable development and societal impact at Asian Business and Management. *Asian Business & Management, 23*(1), 1–8. https://doi.org/10.1057/s41291-024-00262-9

Liu, Y., & Froese, F. J. (2020). Crisis management, global challenges, and sustainable development from an Asian perspective. *Asian Business & Management, 19*(3), 271.

Mahbubani, K. (2022). On the dawn of the Asian century. In K. Mahbubani (Ed.), *The Asian 21st century* (pp. 73–76). Springer Singapore. https://doi.org/10.1007/978-981-16-6811-1_13

Mok, K. H. (2008). Varieties of regulatory regimes in Asia: The liberalization of the higher education market and changing governance in Hong Kong, Singapore and Malaysia. *The Pacific Review, 21*(2), 147–170. https://doi.org/10.1080/09512740801990220

Naafs, S., & Skelton, T. (2020). Introduction: 'Youthful futures? Aspirations, education and employment in Asia.' In Realities and aspirations for Asian youth (pp. 1–14). Routledge. https://doi.org/10.4324/9780429265938-1/introduction-suzanne-naafs-tracey-skelton

Pandey, N., Sinha, A. K. & Kumar, A. (2025). A systematic literature review of extended reality in marketing: An Asian perspective with a focus on India. In M. S. Balaji, J. Bhattacharyya, H. F. Cheong, & W. M. Lim (Eds.), *Future of business in Asia.* Springer.

PwC. (2017). The world in 2050. PwC. Retrieved February 27, 2025, from https://www.pwc.com/gx/en/research-insights/economy/the-world-in-2050.html

Rahman, M. N., Iqbal, B. A., & Rahman, N. (2023). Impact of US-China trade war on Asian economies: Neural network multilayer perceptron approach. *Journal of Chinese Economic and Foreign Trade Studies, 16*(2), 172–189.

Seong, J., Bradley, C., Leung, N., Woetzel, L., Ellingrud, K., Kumar, G., & Wang, P. (2023). Why Asia's economy is on the cusp of a new era. McKinsey. Retrieved February 27, 2025, from https://www.mckinsey.com/mgi/our-research/asia-on-the-cusp-of-a-new-era

Shankar, S. (2024). The opportunities for international expansion in Asia. Retrieved February 26, 2025, from https://amba.ams3.cdn.digitaloceanspaces.cominsights/the-opportunities-for-international-expansion-in-asia

Subramaniam, R., & Ng, T. H. (2014). Beyond factory Asia: Fuelling growth in a changing world. In *Future of factory Asia* (pp. 1–38).

Suhaily, S., Zainuldin, M. H., & Oon, E. Y. N. (2025). Future-proofing business: The changing landscape of internal audit. In M. S. Balaji, J. Bhattacharyya, H. F. Cheong, & W. M. Lim (Eds.), *Future of business in Asia.* Springer.

Suplico-Jeong, L., Guillen, N. B., & Santos, F. R. (2025). Asian tigers and cubs: Shaping the future of business. In M. S. Balaji, J. Bhattacharyya, H. F. Cheong, & W. M. Lim (Eds.), *Future of business in Asia.* Springer.

The Economist. (2019). Where do the Asian tiger economies go from here? The Economist. Retrieved March 3, 2025, from https://www.economist.com/special-report/2019/12/05/where-do-the-asian-tiger-economies-go-from-here

The Economist. (2021). Kishore Mahbubani says this is the dawn of the Asian century. The Economist. Retrieved from February 27, 2025, from https://www.economist.com/the-world-ahead/2020/11/17/kishore-mahbubani-says-this-is-the-dawn-of-the-asian-century

Tonby, O., Woetzel, L., Choi, W., Seong, J., & Wang, D. (2019). Asia's future is now | McKinsey. Retrieved March 1, 2025, from https://www.mckinsey.com/featured-insights/asia-pacific/asias-future-is-now

Tran, A. T., Nguyen, A. T., & Nguyen, T. H. (2025). Towards a sustainable business model in light of inclusiveness, circularity, and green business practices—A conceptualization. In M. S. Balaji, J. Bhattacharyya, H. F. Cheong, & W. M. Lim (Eds.), *Future of business in Asia.* Springer.

Tyagi, P. (2024, September 6). Here are 10 fastest-growing major economies in the last 10 years. Business Today. Retrieved February 27, 2025, from https://www.businesstoday.in/latest/economy/story/here-are-10-fastest-growing-major-economies-in-the-last-10-years-444722-2024-09-06

Warner, M. (2012). Managing across diverse cultures in East Asia. Routledge. Retrieved March 1, 2025, from https://doi.org/10.4324/9780203106952&type=googlepdf

Wilson, D., Purushothaman, R., & Goldman, S. (2003). *Dreaming with BRICs: The path to 2050* (Vol. 99). Goldman, Sachs & Company, 405.

World Economics. (2024). The Future is Asian: World Economics. Retrieved March 1, 2025, from https://www.worldeconomics.com/Thoughts/The-Future-is-Asian.aspx

Wu, W. W. (2025). Future of shipping in Asia: China and Singapore's coping mechanisms in supply chain. In M. S. Balaji, J. Bhattacharyya, H. F. Cheong, & W. M. Lim (Eds.), *Future of business in Asia.* Springer.

Yang, C. Y. (1984). Demystifying Japanese Management Practices. Retrieved March 2, 2025, from https://hbr.org/1984/11/demystifying-japanese-management-practices

Yergin, D., & Stanislaw, J. (2002). The commanding heights: The battle for the world economy. Simon and Schuster. Retrieved March 1, 2025, from https://books.google.com/books?hl=en&lr=&id=uNYzPUhXhJYC&oi=fnd&pg=PA1&dq=Yergin+and+Stanislaw,+2002&ots=hdsrDciTpI&sig=X-BpYyd1OhP7mfJpi8wOveUPtQU

M. S. Balaji is an Associate Professor of Marketing at RMIT University, Australia. Dr. Balaji's research interests include sustainability marketing, retail marketing, and services marketing. He has published over 90 peer-reviewed articles in journals such as Journal of Retailing, Tourism Management, Journal of Travel Research. Journal of Business Research, European Journal of Marketing, International Journal of Hospitality Management, Technological Forecasting and Social Change among others. He serves on editorial board of several journals including International Journal of Contemporary Hospitality Management, Journal of Service Theory and Practice, Services Marketing Quarterly, *International* Journal of Advertising and Journal of Research in Interactive Marketing.

Jishnu Bhattacharyya is a Lecturer (Research) in Marketing at Swansea University's School of Management, United Kingdom. He obtained his Ph.D. in Marketing from the University of Nottingham and previously held the position of Project Scientist at the Indian Institute of Technology Delhi. Dr. Bhattacharyya's research portfolio encompasses a broad range of interconnected domains, including sustainability communication, behaviour change and sustainability, digital marketing, human–robot interaction (service failures), and the psychology of human–AI interactions. His scholarly endeavours are driven by a commitment to integrating rigorous research with pedagogical practice and leveraging the discipline of marketing to foster positive societal change. He is a staunch advocate for data-centric, analytics-driven, and evidence-based approaches to marketing. His research has been published in leading journals such as the Journal of Travel Research and the Journal of Business Research.

Part I
Foundations and Future Perspectives

Chapter 2
Asian Tigers and Cubs: Shaping the Future of Business

Clusters of Future Studies: Corporate Foresight; Past and Futures; Post-normality and Complexity; Technological Trends

Luz Suplico-Jeong, Nelson B. Guillen Jr., and Francia R. Santos

Abstract Asia's rapid economic development offers critical insights into the future of business amidst growing global complexity. The "Asian Tigers"—Hong Kong, South Korea, Taiwan, and Singapore—exhibited high economic growth rates and achieved remarkable growth through industrialisation and export-driven strategies from the 1960s onwards, laying a foundation for economic dynamism. More recently, the "Asian Cubs"—Indonesia, Malaysia, the Philippines, Thailand, and Vietnam—replicated similar economic performances to the economic performances of Asian Tigers have emerged as engines of growth, following in the Tigers' footsteps but with unique adaptations. This systematic review explores how corporate foresight, public–private partnerships, and sustainability initiatives in these economies contribute to structural transformation and resilience. While successes offer replicable lessons, challenges such as inequality, climate change, and technological disruption demand adaptive policy models. By synthesising Asia's symbiotic government-industry frameworks, this analysis provides a reference for managing 21st-century complexities and balancing economic and social welfare.

Keywords Asia · Economic development · Corporate foresight · Tiger economies · Sustainability impacts · Future of business

---

L. Suplico-Jeong · N. B. Guillen Jr. · F. R. Santos (✉)
Department of Marketing and Advertising, Ramon V. del Rosario College of Business, De La Salle University, Manila, Philippines
e-mail: francia.santos@dlsu.edu.ph

L. Suplico-Jeong
e-mail: luz.suplico@dlsu.edu.ph

N. B. Guillen Jr.
e-mail: nelson.guillen@dlsu.edu.ph

M. S. Balaji et al. (eds.), *Future of Business in Asia*,
https://doi.org/10.1007/978-3-031-95940-0_2

2.1 Introduction

Asia's remarkable economic ascent has positioned the region as a pivotal player in the global economy, with growth patterns that offer valuable lessons in business and development. The "Asian Tigers", a term referring to Hong Kong, Taiwan, South Korea, and Singapore, serve as early exemplars of rapid industrialisation, driven by export-led growth and strong state-industry coordination (Mandzieva et al., 2018). These economies achieved high economic growth rates from the 1960s onwards, establishing themselves as key contributors to global business and trade. In recent decades, countries like Indonesia, Malaysia, the Philippines, Thailand, and Vietnam, collectively known as the "Asian Cubs", have started to replicate similar growth trajectories, adapting the Tiger model to their unique political and economic contexts (Chang, 2022).

Figure 2.1 shows Southeast Asian countries: Indonesia, Malaysia, the Philippines, Thailand, and Vietnam. There are five countries that follow the growth and economic path of the tiger economies (Mandzieva et al., 2018).

Asia's rapid development trajectory, impact on global futures, and policy-business interdependencies represent expansive research domains encompassing numerous disciplines. This chapter of the book synthesises evidence across fundamental questions, although additional scholarly perspectives offer deeper understanding, particularly the lessons from Asia's development models, highlighting policy-business interdependencies and their potential application to other developing regions. As countries worldwide confront rising inequalities, climate instability, and technological disruption, the Asian Tigers and Cubs offer a framework for sustainable growth that is both equitable and resilient.

Fig. 2.1 Tiger cub economies in Southeast Asia. *Photo/Image Source* Google Maps of Asia; Flag images from Canva

On a country's progress models, Ohno (2009) dissects industrial revitalisation strategies, exploring how developing economies transition from resource exporters to innovation hubs. Yusuf (2022) specifies the innovation system capabilities that power leading East Asian economies through various phases. Critically examining catalysts and constraints within these maturing country cases helps inform sustainable transitions.

Meanwhile, complex systemic resilience and risk readiness have emerged as essential governance challenges of the 2020s, prompting research on supply chain finance networks (Chew et al., 2022), health infrastructure (Zhang et al., 2022), and other areas. Ongoing conceptual and empirical efforts to quantify resilience factors and simulate policy responses could better equip societies for turbulent times ahead.

Business sustainability also draws increasing attention just like how the Asian product-service system models have been balancing environmental sustainability with commercial viability, as described by Chang and Chen (2022). Regional studies, such as Thuzar's (2021) assessment of Southeast Asian green finance, reveal notable trends and gaps. Expanding sustainability research to encompass social enterprise and nonprofit approaches could yield valuable cross-sectoral insights.

This chapter aims to analyse the mechanisms that facilitated Asia's development trajectories and assess their relevance in addressing contemporary global challenges. Concepts like *past cubs* and *corporate foresight* are pivotal in understanding how Asian Economies sustain growth amidst change, with the latter defined as the strategic capability to anticipate and shape future opportunities (Rohrbeck & Kum, 2018). Additionally, the integration of sustainability impacts into development strategies underscores the importance of balancing rapid growth with environmental and social welfare.

2.1.1 Corporate Foresight

Corporate foresight refers to the ability of companies to analyse current trends and develop informed visions of the future to guide strategic planning (Rohrbeck & Kum, 2018). This capability enables leading firms to anticipate and capitalise on emerging opportunities, a role exemplified by Asian tiger and cub economies, which highlight corporate foresight as crucial for national development.

Many well-known Asian corporations that spurred significant economic growth engaged in long-range planning and forecasting. For instance, Samsung focused its investments on memory chips and display panels well before its competitors, facilitating substantial expansion (Jun & Suh, 2013). By monitoring technological advances and shifts in global consumer demand, Samsung gained first-mover advantages. Likewise, strategic investments in electric vehicles and batteries underpin contemporary moves by companies like BYD Auto and LG Chem (Peterson et al., 2022).

Beyond individual products and technologies, corporate foresight in Asia also extends to entirely new sectors with promising futures. Indonesian unicorn GoTo, for instance, preemptively built infrastructure for online payments and deliveries, facilitating rapid growth (Dalton, 2022). By envisioning digital lifestyles before they were fully embraced by consumers, GoTo accelerated adoption as e-commerce capabilities matured. Such private sector initiatives in corporate foresight contribute to broader economic progress.

Governments across Asia now actively collaborate with industry leaders on national foresight projects to align development roadmaps (Rohrbeck & Kum, 2018). The incorporation of corporate perspectives ensures that resulting policies remain responsive to commercial realities, particularly in areas such as sustainability, smart cities, and global integration. This public–private foresight approach fosters innovation ecosystems at the local level.

2.1.2 Past Cubs

The "Asian Cubs" refer to the emerging economies of Southeast Asia—Indonesia, Malaysia, the Philippines, Thailand, and Vietnam—that have shown rapid growth (Chang, 2022). While the mature "Asian Tiger" economies of Singapore, Hong Kong, Taiwan, and South Korea achieved urbanisation between the 1960s and 1990s, the Cubs represent a subsequent wave of rising Asia–Pacific powers.

Analysing their historical developmental trajectories provides valuable insights. Generally, the Asian cubs have pursued export-orientated growth by leveraging comparative advantages in agriculture and labour-intensive manufacturing (Studwell, 2020). Revenues from commodity exports and integrated roles in global supply chains provided foreign reserves, which were reinvested in infrastructure and human capital. However, these economies varied in their approaches to state-business relations and domestic institution-building, resulting in significant differences in outcomes.

Figure 2.2 shows that in the 1950s, GDP per capita income "ranged from $820 in Indonesia and Thailand and $1560 in Malaysia. By 2010, income levels ranged from $3020 in the Philippines to $10,090 in Malaysia." Over the last 60 years, these countries have developed significantly, becoming influential players in the global economy. The income levels of these "Asian Cubs"—Thailand, the Philippines, and Indonesia—were close to those of the Asian Tigers in the 1950s, as depicted in Fig. 2.2 (Restrepo-Echavarría & Arias, 2017).

In the ASEAN region, Malaysia, Thailand, the Philippines, and Indonesia are known as the "Tiger Cub Economies" Although they developed less rapidly than the original Tigers, they grew at a more consistent rate over this period.

Vietnam's authoritarian governance facilitated disciplined market reforms, which attracted foreign direct investment, boosted productivity, and catalysed structural transformation (Malesky, 2022). In contrast, more democratically accountable countries like Indonesia and the Philippines struggled with vested interests that hindered industrial policy effectiveness. Thailand's hybrid model, balancing cooperation with regulatory control, fostered relatively inclusive development (Brimble et al., 2022).

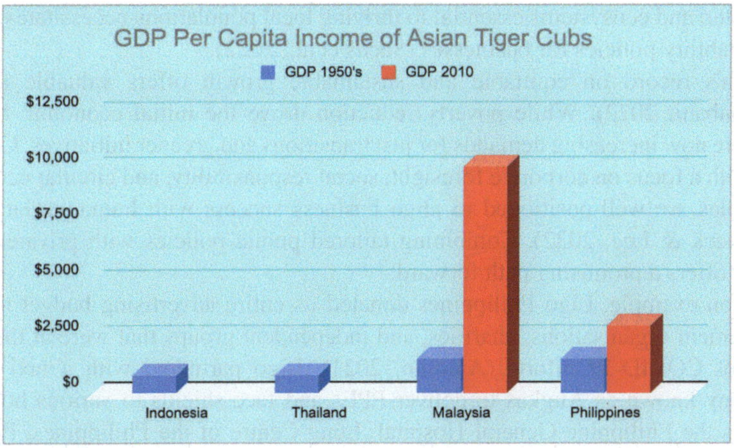

Fig. 2.2 GDP per capita income of Asian tiger cubs. *Source* Restrepo-Echavarria and Arias (2017)

Malaysia's initial push into heavy industry eventually stalled due to resource misallocation and inadequate discipline (Rasiah et al., 2022). Subsequent policy shifts, intended to move the economy up the value chain, increasingly relied on importing skilled labour rather than developing local talent. Such differences underscore how institutional structures influence sustainable development trajectories.

2.1.3 Sustainable Futures of the Business and Humanity

Asia's rapid development carries substantial implications for the future of both commerce and humanity. Business drives material advancement central to human well-being, while commercial motives shape emerging technologies and lifestyles. Reviewing the trajectories of the Asian Tigers and Cubs reveals models for managing this complex relationship sustainably.

On its current trajectory, Asia is expected to account for more than half of global GDP by 2050, with China and India projected to become the world's largest and third-largest economies, respectively (PwC, 2022). New consumer classes across Asia will drive demand in markets from energy to housing. Corporate strategies today are increasingly focused on anticipating these shifts, while policies prioritise leadership in renewable energy, electric vehicles, and automation to meet future demand (Rohrbeck & Kum, 2018).

However, the hyper-growth associated with this development also brings risks, including inequality and environmental degradation. Addressing these issues requires balancing profits with societal well-being (Stiglitz, 2022). Likewise, protecting clean air, water, and ecosystems essential to thriving local populations necessitates robust sustainability policies for businesses (Sachs et al., 2022).

Asia's record on equitable and sustainable growth offers valuable lessons (Mahbubani, 2022). While poverty reduction drove the initial economic booms, there are now increasing demands for just transitions and greener initiatives. Companies with a focus on corporate foresight, social responsibility, and circular economy principles are well-positioned to align business success with humanitarian goals (Rohrbeck & Lee, 2022). Combining tailored public policies with private sector efforts offers a promising path forward.

As an example, Ligo Philippines donated its entire advertising budget to non-government organisations, charities, and independent groups that were at the forefront of COVID-19 efforts (Antonio, 2021). Ligo partnered with a taxi motor company known as Angkas to deliver PPEs and face shields to various hospitals such as the Philippine General Hospital, Lung Centre of the Philippines, Premier Medical Centre, and Ospital ng Muntinlupa. Ligo Philippines has also been active in humorous social media posts on reminding Filipinos to social distancing and observe the COVID-19 protocol (Cirineo, 2020). After COVID-19, Ligo Philippines continued with its social media posts that displayed its care and concern for Filipinos.

2.1.4 Impacts of the Business and Environmental Futures

Asia's rapid economic growth has had significant environmental consequences, highlighting the dual nature of development. The extensive resource extraction, high-pollution production, and unchecked consumption fuelling regional booms now pose threats to ecological stability (Sachs et al., 2022). However, the same technological capabilities and financial resources that enabled rapid industrialisation can also support sustainable solutions.

Asia's environmental impact has grown substantially in absolute terms but remains lower than Western levels on a per capita basis (ADB, 2020). Yet, without corrective measures, emissions in Asia will likely exceed scientifically established limits. Likewise, biodiversity losses and pollution in Asia, while quantitatively smaller, reflect patterns seen in Western industrial development (WWF, 2020). Key drivers include fossil fuel-heavy Chinese manufacturing and Indonesian palm oil cultivation.

Leading companies and governments in Asia increasingly acknowledge ecological limits. Many firms are adopting renewable energy, electrifying fleets, improving efficiency, and implementing circular design principles in response to social pressures and regulatory changes (Rohrbeck & Lee, 2022). Countries like Singapore, Japan, and South Korea lead in corporate sustainability, though progress remains limited in nations with weaker state capacities.

Ultimately, Asia both threatens global environmental stability and offers innovative pathways for green transitions. Integrating local public policies with private initiatives around decarbonisation and conservation can balance continued human development with ecological integrity (Sachs et al., 2022). International technology transfers and cooperation can further facilitate sustainable trajectories.

2.1.5 *Post-pandemic Normality and Complexity*

The COVID-19 pandemic severely disrupted Asia's economic norms but also exposed underlying vulnerabilities and adaptive responses for the future. The current focus has shifted toward building resilient systems capable of managing greater post-pandemic complexity (Asian Development Bank, 2021). Key priorities include artificial intelligence, digitalisation, policy coordination, and supply chain restructuring.

The pandemic underscored Asia's economic interconnectedness and the cross-border impacts of global crises (Mahbubani, 2022). Research on COVID-19's impact highlights the importance of developing resilient systems rooted in communities' reactive and proactive capacities, as seen in Japan and the Philippines. In business, leadership is essential for successfully navigating future threats and uncertainties (Santos & Yamada, 2023). Technological investments that enabled remote work and e-commerce buffered the impact of the pandemic while underscoring the potential of digital transformation. Consequently, governments are prioritising connectivity upgrades, data systems, and smart infrastructure to enable real-time monitoring, agile policy responses, and transparent information flows (Camba, 2022).

The pandemic also revealed disparities in response capabilities and the need for greater coordination. While Northeast Asia leveraged reserves and effective containment measures, Southeast Asia faced difficulties due to informal economies, limited funds, and weak healthcare infrastructure (Asian Development Bank, 2021). New governance frameworks, public–private partnerships, and regional cooperation mechanisms aim to better handle future crises.

Lastly, the over-reliance on distant links in global supply chains has sparked re-evaluations, with calls for more localised production, inventory buffers, and diversified suppliers to enhance resilience. However, de-globalisation poses risks of reduced efficiency and increased complexity (Strange, 2020). Navigating the post-COVID landscape will require a careful balance of efficiency, resilience, and cross-border collaboration. Asia's adaptive responses offer models for shaping global policies and corporate strategies.

2.1.6 *Business and Technological Trends*

Rapid development across Asia is closely linked with the commercial adoption of emerging technologies. Competitive pressures and large markets drive massive investments in digital infrastructure, automation, artificial intelligence, biotechnology, and green technologies, which help diffuse innovations globally (Xiaolan et al., 2021). However, these trends bring new risks that require effective mitigation.

Asian companies are heavily investing in next-generation research and development, covering areas from semiconductors to electric vehicles and solar photovoltaic (PV) systems and consistently securing world-leading patent totals in sectors anticipated to transform business and society (World Intellectual Property Organisation, 2022). Government industrial policies bolster these efforts by prioritising sectors vital for economic or military competitiveness, as seen in Taiwan's semiconductor industry and China's expertise in networking technology.

E-commerce has become essential in the global retail landscape, with many firms experiencing substantial transformations due to the internet and ongoing digitalisation. There are now over five billion internet users globally, with numbers continuing to rise, which translates into more consumers shopping online (Statista, 2023). Technological advancements significantly impact profitability and the future of business in an increasingly interdisciplinary world (Chaudhari, 2023). While the exact trajectory of technological transformation remains uncertain, these trends will undoubtedly affect work and living standards worldwide. As such, effective coordination across governments, educational institutions, businesses, and civil society is essential (Chaudhari, 2023). However, technology and digitalisation races also heighten trade disputes and security concerns (Mahbubani, 2022). Developments in surveillance, automation, synthetic media, and the debatable artificial intelligence simultaneously present both advancements and threats. Ensuring equitable access and establishing ethical guidelines amid rapid technological change is complex yet essential.

Ultimately, Asia is poised to lead numerous emerging business and technology revolutions this century, powered by its large commercial capacity and state support. However, global cooperation around governance and prioritising innovations aligned with human welfare will be crucial to maximising social benefits while mitigating risks.

Corporate foresight capabilities in firms like Samsung and GoTo aligned business success with national objectives through long-term investments (Rohrbeck & Kum, 2018). Export-led industrial policies have further leveraged globalisation, although the quality of collaboration affects outcomes, as seen in the contrasting cases of Indonesia and Vietnam (Studwell, 2020).

Rapid growth now presents ecological threats but also encourages green transitions driven by corporate leaders and supported by regulatory frameworks (Sachs et al., 2022). Globalisation has fostered a learning marketplace, increasing demand for international communication, e-learning, and artificial intelligence (Azoth Analytics, 2023).

The complexities introduced by the post-pandemic world demand coordination across government, industry, academia, and civil society to strengthen systemic resilience (Mahbubani, 2022). Key priorities for businesses include digitisation, supply chain buffers, and employee skill development.

Asia's vast markets and competitive pressures fuel technological innovation in sectors ranging from semiconductors to electric vehicles, leading to global patents and raising ethical considerations (World Intellectual Property Organisation, 2022). Managing these complex innovation ecosystems sustainably calls for updated state-business models that balance economic prosperity with social welfare (Chang, 2022).

Applying these insights globally remains essential. Toolkits derived from Asia's lessons can aid developing countries in implementing reforms (Studwell, 2020), and Asian corporations' documentation of their practices could help improve governance in technology and business worldwide (Deloitte, 2022). Continued research bridging economic, social, environmental, and technological perspectives is essential to progress.

Latest study of the World Bank (2024) reveals that new business models have emerged along with new technologies such as digitisation and digitalisation platforms. These technologies enhance the efficiencies of the business, in general. In 2023, the size of the digital platform economy contributed 5–7% of GDP in most East Asia Pacific countries. These platforms give entrepreneurs access to small business opportunities in the services sector (2024).

While this review distils the current interdisciplinary knowledge on Asian development, significant research opportunities around sustainable business, complex systems management, and governance innovations remain to be explored. Integrating economic, social, technological, and environmental perspectives from both global South and North perspectives can deepen the discourse. Continued scholarly exploration of the dynamics highlighted here will further advance equitable futures worldwide.

2.2 Review Methodology

This study employs a Systematic Literature Review (SLR) to analyse Asia's development trajectories and their implications for sustainable global progress. Following rigorous systematic review protocols (Paul et al., 2021), this methodology ensures a structured, transparent, and replicable approach in identifying, evaluating, and synthesising scholarly literature across diverse disciplines. These disciplines include development economics, business strategy, sustainability, and futures studies. Using SLR, the study pulls out important lessons from the experiences of the Asian Tigers and Asian Cubs to show off their strategic economic models and see how they can be used around the world to create sustainable futures.

The SLR process integrates Preferred Reporting Items for Systematic Reviews and Meta-Analyses (PRISMA) guidelines (Moher et al., 2009) to enhance methodological rigour and replicability. This ensures a comprehensive assessment of scholarly contributions while minimising selection biases and methodological inconsistencies. The review aligns with SPAR-4-SLR (Scientific Procedures and Rationales for Systematic Literature Reviews) standards (Paul et al., 2021), providing a robust framework for organising literature synthesis and thematic categorisation.

2.2.1 Literature Search Strategy

The literature search was conducted across multiple electronic academic databases—JSTOR, ProQuest, Scopus, and Web of Science—to ensure comprehensive coverage of relevant fields up to the current year. Search terms combined keywords related to Asian economic development and future-orientated business models, including "Asian Tigers," "Asian Cubs," "corporate foresight," "sustainability impacts," and "business strategy." The initial search yielded only 175 articles from 2010 to 2024, of which 40 met the inclusion criteria based on their relevance to Asia's development models and implications for sustainable growth. To enhance the robustness of the review, supplementary searches were conducted using Google Scholar and industry-specific reports, ensuring that significant grey literature and policy documents were included when relevant. This multi-source approach mitigated database biases and improved the depth of insights gathered.

Online searches were conducted until the same articles and books repeatedly appeared, even when different search terms and techniques were employed. Efforts were made to ensure an exhaustive search process. Relevant and contemporary articles were also searched and added.

2.2.2 Inclusion and Exclusion Criteria

The data visualisation of the literature search is shown in Fig. 2.3. The inclusion and exclusion criteria are discussed in Sect. 2.2.2. To further validate the selection process, citation tracking, and snowballing techniques were utilised to identify influential works cited in high-impact papers. This approach ensured that seminal contributions and foundational theories relevant to Asia's development were captured. Additionally, inter-rater reliability checks were conducted wherein independent researchers reviewed selected papers to assess consistency in inclusion decisions, thereby minimising potential biases in study selection.

Inclusion criteria focused on English-language articles addressing:

1. The development trajectories of Asian Tigers and Cubs,
2. Corporate foresight in business strategy,
3. Sustainability impacts within Asian economic contexts, and
4. Policy-business interdependencies in Asia.

Articles that lacked regional or thematic relevance or did not contribute substantially to the systematic review framework were excluded. The selection process prioritised peer-reviewed journal articles and academic book chapters; significant policy papers and reports were also included to provide a broader perspective.

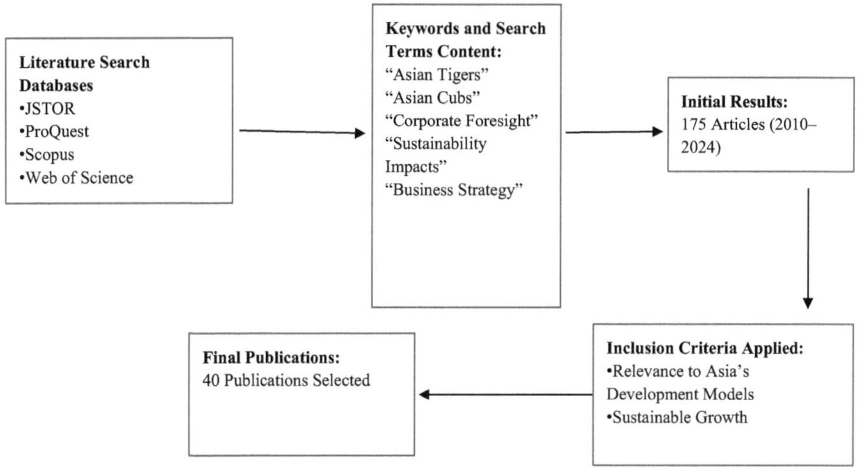

Fig. 2.3 Data visualisation of the literature search

2.2.3 Data Extraction and Coding

Following the article selection, data extraction involved identifying key themes, results, and methodological approaches within each publication. Deductive coding was used to categorise evidence into five primary areas: national development strategies, corporate foresight, environmental impacts, technological advancements, and policy responses. Additionally, supplementary inductive coding captured emergent themes, such as public–private collaboration, complex systems resilience, and lessons transferable to other regions. This method made it possible to do an integrated analysis that combined structured overviews with broad insights on paths for sustainable development.

Figure 2.4 presents data extraction and a framework for coding. The systematic review of literature started with 175 articles as initial search hits. Subsequently, the search was narrowed down to 40 publications from 2010 to 2024. A systematic review of literature outcomes highlights the organisation of themes using deductive and inductive approaches.

2.2.4 Limitations

This study's limitations include a predominant focus on English-language publications, which may limit insights from non-English regional sources. Furthermore, inherent challenges in comparing data across diverse methodologies and disciplinary perspectives persist, though the multidisciplinary approach aims to counterbalance specific blind spots. Despite these limitations, this SLR thoroughly synthesises Asia's

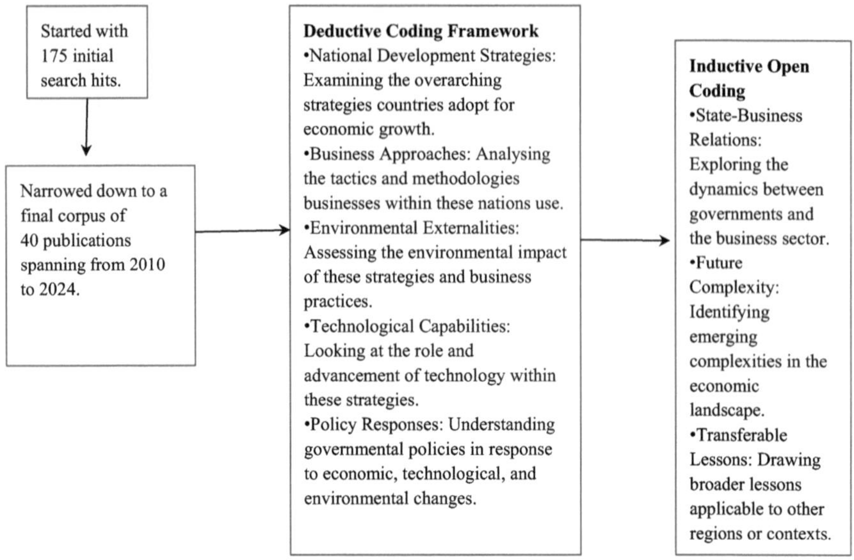

Fig. 2.4 Data extraction and coding. *Note* This figure is the authors' original narrative

development models, positioning them within the broader discourse on sustainable economic progress.

While traversing all possibilities of literature review inclusions, the study was able to cover relevant emerging themes and profound insights introduced in the findings of this chapter. This review focused more on strategic insights for the businesses to become future-ready and economically resilient. However, a separate chapter should highlight clusters of future studies, such as "humanity and environmental futures" and green transitions in business. This review targeted clusters of future studies such as corporate foresight, past and futures, post-normality and complexity, and technological trends.

2.3 Findings

The literature analysis summarises key themes and insights, presents results that align with the systematic review's thematic structure, and emphasises patterns and implications drawn from Asia's economic trajectory.

The review of the five (5) emerging themes: National Development Strategies; Corporate Foresight and Strategic Planning; Sustainability and Environmental Impacts; Technological Advancements and Digital Transformation; and Policy-Business Interdependencies demonstrated the shared value of Asia's Tiger and Cub economies' implications for transformative pathways to global development trajectories *i.e.,* synergies in state-business relationships, corporate foresight as a strategic tool, sustainable business models that respond to environmental challenges, technology and digital transformation as a catalyst for growth, and policy recommendations for future resilience and global impact. Profound insights from the Asian Tigers and Cubs provide a robust foundation for equitable, resilient, and future-ready business strategies across the globe.

Learning insights from the Asian Tigers and Cubs did not only produce emerging themes but provided clarity of foundations for countries across the globe to prepare for whatever future uncertainties they may experience. Key findings include the following:

2.3.1 National Development Strategies

Studies (Ohno, 2009; Studwell, 2020) highlight how the export-oriented models of the Tigers and Cubs facilitated rapid industrialisation and economic gains. These strategies underscore state-led initiatives to foster industrial clusters, support targeted sectors, and attract foreign direct investment.

Dow Jones Institutional News (2024) reported that companies of well-known brands looked beyond the Southeast Asian countries to diversify the manufacturing of products out of China. Apple has just chosen India for manufacturing millions of iPhones, while Nike decided to settle in Vietnam for its famous sneakers.

Vietnam has become the world's second largest smartphone exporter next to China because of producing about one third of Samsung Electronics' products (Business Sweden, n.d.). Philippines and Indonesia must also experience the export-led manufacturing that boomed Asian Tigers like South Korea and Taiwan.

2.3.2 Corporate Foresight and Strategic Planning

There is evidence from big Asian companies (Jun & Suh, 2013; Rohrbeck & Kum, 2018) that foresight helps companies lead market changes, especially in the tech and digital fields. Corporations like Samsung and GoTo leveraged foresight capabilities to proactively address consumer needs and invest in emerging sectors.

Strategic planning needs a comprehensive analysis of the industry in both micro and macro-economic perspectives. For instance, the industry 6.0 ecosystem transforms people's culture, and the work of humans becomes highly efficient with the utilisation of AI (Chaudhari, 2023). Consequently, the future of the industry must be secured, supported and anticipated through corporate foresights that advance strategies for sustainable development and resilience.

2.3.3 Sustainability and Environmental Impacts

Asian economies display diverse approaches to sustainability. Tigers such as Singapore and South Korea implement stringent policies, while other countries face challenges due to lower regulatory capacities (Sachs et al., 2022). Key practices include adopting renewable energy and circular economic models.

Malaysia and the Philippines are two Asian economies that will sustain performance in terms of a stronger domestic demand. Based on the analysis of Nomura (2024), Malaysia's construction boom, shifts in the supply chain and public infrastructure developments are contributing to the economy. The Philippines, on the other hand, gains from infrastructure expenditures as the election in May 2025 approaches.

A sustainable business model must be adopted in the volatile, uncertain, complex, and ambiguous (VUCA) environment. Hence, it is necessary to devise sustainable strategies such as "product stewardship, clean technology, pollution prevention, and sustainability vision. To enable this, sustainability awareness plays a key role (Karuppiah et al., 2023).

2.3.4 Technological Advancements and Digital Transformation

Asia's rapid adoption of technology, driven by competitive markets and government support, has positioned it as a leader in fields like AI, automation, and green technologies.

The Southeast Asian region transitioned quickly into a mobile economy, as reported by PR Newswire (2016), with Singapore leading the economy in the region. Indonesia is ranked #4 as a nation with the largest population globally. Experts projected a significant penetration of smartphones in Vietnam and the Philippines.

There will be a great demand for AI. Studies highlight the link between investments in digital infrastructure and resilience, particularly in the post-COVID-19 landscape (Asian Development Bank, 2021). The International Monetary Fund (IMF) predicts that 40% of all jobs will be impacted by AI, and this may lead to rising inequality. The integration of AI, however, can help workers in their productivity enhancement (Liang, 2024).

2.3.5 Policy-Business Interdependencies

The synergy between the state and business sectors in the region illustrates a model in which public and private sectors co-evolve, particularly in urban development, the digital economy, and green infrastructure (Camba, 2022). This interplay supports Asia's ongoing competitiveness within an increasingly complex global landscape.

The growth of the middle class and improved internet infrastructure led to the ASEAN region's fast-changing mobile economy in the 2020s (PR Newswire, 2016). The experiences of Asia's Tiger and Cub economies illustrate transformative pathways for developing sustainable and resilient business models in a rapidly evolving global context. These economies show how well-coordinated state-business frameworks, corporate foresight, and policies that focus on sustainability can work. Each of these has different effects on the development paths of Asia and the world. In summary, the trajectories of the Asian Tigers and Cubs reveal a catalytic state-business synergy that promotes equitable growth but faces sustainability challenges amid growing complexity.

Further examination of the Asian economic model reveals nuanced adaptations across different national contexts, with varying degrees of state intervention and market liberalisation creating distinct development pathways. Research by Kimura and Chen (2018) demonstrates how these economies have strategically balanced industrial policy with market mechanisms, creating dynamic comparative advantages through targeted education investments, selective protectionism, and aggressive export promotion. The region's success in creating innovation ecosystems stems from this pragmatic flexibility—allowing for strategic state guidance while maintaining sufficient market competition to drive productivity gains. One interesting thing about how digital transformation has sped up institutional evolution is how regulatory sandboxes and experimental policy zones have made it possible for governance systems to change faster than normal legislative processes allow. This ability of institutions to change quickly has become a competitive advantage in and of itself, letting these economies respond quickly to technological disruptions and global economic shocks while keeping society together during times of structural change.

2.3.6 *Synergies in State-Business Relationships*

The intertwined nature of government and industry in Asian development models has fostered a unique synergy that supports economic resilience and technological leadership. In economies like Singapore and South Korea, government-led initiatives align closely with business objectives, from infrastructure investments to export-orientated growth policies (Studwell, 2020). This state-business synthesis enables these countries to respond dynamically to crises, such as the COVID-19 pandemic, by leveraging centralised planning and agile corporate support (Mahbubani, 2022).

Interdependence in East Asia regional integration—the notable ASEAN region + 3—gained attention and showed increasing significance in the world economy (Rui & Yao, 2016). For other regions, the potential benefits of public–private collaboration in designing long-term development strategies—particularly in sectors like digital infrastructure and green technology—are evident. Policymakers worldwide may find value in adopting similar state-industry coordination to enhance resilience in critical sectors like healthcare, technology, energy, and infrastructure development.

A study from the Asian Development Bank (Eggleston, 2023) showed synergies in state-business relationships and researchers have been used to contain COVID-19. The study has pointed out that several countries have tapped public–private partnerships to achieve the Social Development Goals. One of these countries is the Philippines where national plans have been developed beyond disease-specific programs to tap private sector collaboration in healthcare (Eggleston, 2023). Beyond healthcare, there is also a growing popularity of Public–Private Partnerships (PPPs) in the context of infrastructure development (Endo & Ram, 2021). In the Philippines, examples of PPPs include the Cavite-Laguna Expressway, North Luzon Expressway-South Luzon Expressway Connector Project and the Tarlac-Pangasinan-Union Expressway which have considerably shortened travel from these provinces to Metro Manila and vice versa.

2.3.7 Corporate Foresight as a Strategic Tool

Corporate foresight as demonstrated by firms like Samsung and BYD, is integral to Asia's competitive advantage, particularly in technology sectors. By anticipating global trends in electronics, renewable energy, and digital transformation, these firms maintain market dominance and drive advancements across their sectors (Rohrbeck & Kum, 2018). This strategic foresight shows the potential for firms worldwide to engage in proactive planning to navigate disruptions and capture growth opportunities.

Globally, businesses can leverage foresight and scenario planning to identify emerging trends and proactively respond to shifts in consumer demand, regulatory landscapes, and environmental pressures. This approach could help mitigate risks from sudden technological or societal changes and increase adaptability across various sectors.

2.3.8 Sustainable Business Models in Response to Environmental Challenges

As Asian economies grapple with the environmental consequences of rapid industrialisation, sustainability has become central to policy and corporate agendas. Nations like South Korea and Japan lead the region in corporate sustainability practices, such as adopting renewable energy, circular economy principles, and energy-efficient production methods (Sachs et al., 2022). This emphasis on environmental stewardship positions Asia as a testing ground for integrating economic growth with ecological balance.

These practices indicate that sustainable business models—emphasising green technologies, resource efficiency, and pollution control—are essential for mitigating climate impacts globally. Regions facing similar development pressures can look to Asia's evolving sustainability strategies as a model to balance growth with ecological responsibility, aligning with the Sustainable Development Goals (SDGs).

It is crucial for businesses to develop a robust strategy for a successful sustainable business model (SBM). Through a sustainable business model, a superior social and environmental goals plus enhanced economic values can be achieved (Karuppiah et al., 2023).

2.3.9 Technology and Digital Transformation as a Catalyst for Growth

Asia's rapid adoption of technology, driven by both private investment and public policy, has established the region as a leader in digital transformation and automation. Competitive pressures and large markets have catalysed advancements in artificial intelligence, robotics, and green technology with significant ripple effects on global markets (World Intellectual Property Organisation, 2022). Investments in digital infrastructure during the COVID-19 pandemic underscored the crucial role of technology in fostering resilience and growth (Asian Development Bank, 2021). Productivity gains from automated production are increasing its value. Robot adoption, AI, and digitisation (World Bank, 2024) are more likely to be prevalent, especially in ageing country tigers like Japan, South Korea, Singapore, and Taiwan. Hence, businesses are obliged to invest in digital workplaces and use digital technologies and platforms.

For developing regions, prioritising digital infrastructure and innovation can enhance both economic resilience and global competitiveness. As industries increasingly digitalise, incorporating digital capabilities into public services, healthcare, and education offers an effective means to bridge gaps in productivity and quality of life. Robust growth can be projected if not disrupted by any uncertainties and climate change.

Indeed, the integration of emerging technologies such as blockchain, quantum computing, and advanced biotechnology is further accelerating Asia's technological revolution, creating new economic niches and reshaping traditional industries. These cutting-edge innovations are enabling novel business models while simultaneously addressing persistent regional challenges in financial inclusion, healthcare accessibility, and environmental sustainability. Countries like Vietnam, Indonesia, and the Philippines are increasingly positioning themselves as the next wave of digital economies, leveraging their young, tech-savvy populations and strategic investments in digital literacy to climb the value chain in global technology markets. Developing economies skip over the middle stages of technological evolution, which is called "technological leapfrogging." This allows for fast progress but also makes it harder to handle the effects of faster digital transformation on society.

2.4 Future Directions

2.4.1 Future Resilience and Global Impact

Asia's development trajectory highlights the importance of policy innovation to sustain economic growth amid global challenges. For instance, Vietnam has implemented flexible regulatory frameworks that attract foreign direct investment while maintaining control over strategic industries (Malesky, 2022). In contrast, countries with more rigid policies face greater challenges in adapting to rapidly evolving global trends.

The adaptability of policies in response to emerging challenges—such as demographic shifts, climate change, and geopolitical tensions—will be critical for sustainable development. By fostering a collaborative environment where businesses can innovate within regulatory frameworks, other countries may leverage adaptive policy models to promote long-term resilience.

While this review distils the current interdisciplinary knowledge on Asian development, significant research opportunities around sustainable business, complex systems management, and governance innovations remain to be explored. Integrating economic, social, technological, and environmental perspectives from both global South and North perspectives can deepen the discourse. Continued scholarly exploration of the dynamics highlighted here will further advance equitable futures worldwide.

2.4.2 The Rise of Asia's Tiger Cub Economies

The economic rise of Asia's Tiger Cub Economies illustrates a diverse yet interconnected approach to sustainable and resilient growth. These economies have successfully navigated industrialisation and globalisation through coordinated state-industry frameworks, corporate foresight, and sustainability-driven policies—all while addressing modern challenges such as inequality, environmental degradation, and technological disruption. This systematic review highlights how strategic foresight and robust public–private synergies have enabled Asia to balance rapid economic growth with long-term social and environmental welfare.

Asia's development trajectories suggest that these models have global applicability, especially for countries grappling with similar issues of industrialisation, resource constraints, and inequality. In an increasingly complex global landscape, the experiences of the Asian Tigers and Cubs emphasise that collaboration, foresight, and adaptive policymaking are essential for sustaining economic resilience and progress.

By integrating these strategies, other regions and economies can develop robust, future-proof models that align economic goals with social and environmental sustainability. As the world faces pressing issues like climate change, demographic shifts, and digital transformation, the adaptable and innovative approaches of Asia's economies provide a framework for countries seeking to harmonise growth with global responsibilities.

2.5 Conclusion

The findings of this chapter have implications for long-term development, nations' investment decisions, sustainability and business models, transformation and economic resilience, and policy making.

2.5.1 Enhance Public–Private Collaboration for Long-Term Development

Policymakers should foster strategic collaborations between public and private sectors to align business innovation with national development goals. This objective can be achieved by:

- Establishing innovation hubs that bring together industry leaders, academic institutions, and policymakers to address local challenges and foster cross-sector innovation.
- Providing incentives for businesses to engage in sustainable practices, such as tax breaks for green investments or subsidies for renewable energy projects.

2.5.2 Invest in Corporate Foresight and Scenario Planning

Businesses worldwide can adopt corporate foresight practices to anticipate and adapt to future market shifts. To do so, companies should:

- Integrate foresight teams or partner with think tanks to develop strategic planning frameworks based on trend analysis and scenario modelling.
- Regularly update risk management practices to include emerging global challenges, such as climate risks, regulatory changes, and digital disruptions.

2.5.3 Implement Sustainability-Oriented Business Models

Building on Asia's sustainability initiatives, companies and governments can collaborate to promote sustainable growth through:

- Circular economy practices encourage industries to recycle materials, reduce waste, and minimise environmental impacts.
- Emphasising renewable energy adoption, with governments supporting green infrastructure development and companies transitioning to lower carbon emissions in operations.

2.5.4 Prioritise Digital Transformation for Economic Resilience

To enhance competitiveness and resilience, governments and businesses should invest in digital transformation by:

- Developing robust digital infrastructure, including nationwide internet access, data centres, and cybersecurity measures to support digital commerce and services.
- Promoting digital literacy and skills development programs to prepare the workforce for technology-driven industries and boost employability.

2.5.5 Promote Flexible and Adaptive Policy Models

Inspired by the dynamic policies seen in Vietnam and Singapore, policymakers can adopt flexible approaches to address emerging needs, such as:

- Enacting policies that balance economic goals with social welfare, such as affordable housing, healthcare access, and inclusive financial systems.
- Encouraging responsive policy-making processes, where regular assessments allow for quick adjustments to changing conditions in global trade, environmental policy, and technological innovation.

2.5.6 Expand Research on Transferable Insights for Global Application

Scholars and practitioners should build on Asia's lessons by researching how these insights can be applied to other regional contexts. Key areas include:

- Comparative studies of public–private partnerships across different economic regions, particularly in emerging economies with similar developmental stages.
- Exploring how Asia's strategies can inform sustainable practices in industries heavily reliant on natural resources or facing high environmental risks.

2.5.7 *Areas for Future Research*

The exploration of Asian Tigers and Cubs offers several directions for further investigation, particularly as these economies adapt to ongoing global challenges. With challenges like climate change, digital transformation, and rising inequality, businesses and policymakers must leverage innovative practices to maintain economic resilience.

To expand research on transferable insights for global implications, findings of this study may be validated by scholars in other regions or in diverse world contexts. We have identified the following areas for future research to enhance the profound insights gained from studying Asian tigers and cubs. Suggested areas for further research include the following:

2.5.7.1 Comparative Studies on Public–Private Partnerships (PPP)

Future research could conduct comparative analyses of public–private collaborations across regions. Looking into how the different institutional frameworks in Asia, Africa, and Latin America affect innovation and economic resilience can help us come up with flexible partnership models.

In an increasingly interconnected world, the aforementioned economies illustrate how strategic foresight, sustainability, and government-business collaboration can drive prosperity. Thus, comparative studies using the PPP model from diverse perspectives across the globe can offer actionable research and interventions that drive prosperity.

2.5.7.2 Technology Adoption and Digital Transformation

As digital infrastructure becomes increasingly critical, further studies could focus on how emerging economies can accelerate digital adoption. Researchers may look into the role of government incentives, the things that make it difficult to go digital, and the social and economic effects of using automation, AI, and renewable energy.

2.5.7.3 Sustainable Development and Circular Economy Models

Sustainable growth models, such as the circular economy, are gaining traction in Asian policy circles. In the future, researchers could look at how well these models work in places with limited resources. They could focus on certain industries (like manufacturing and agriculture) to find the best ways to balance economic growth with environmental protection.

2.5.7.4 Corporate Foresight and Strategic Resilience

While corporate foresight has proven beneficial in the Asian context, further research could explore its adaptability in different cultural and economic environments. Studies could investigate how companies outside Asia, they might integrate foresight techniques to better prepare for disruptions, including climate risks, technological shifts, and market volatility.

2.5.7.5 Policy Innovation for Socio-economic Equity

Policy flexibility in countries like Vietnam and Singapore demonstrates the potential for responsive governance in rapidly changing environments. Research could examine how adaptive policy models can mitigate inequality and support inclusive growth, focusing on health, education, and economic access in underdeveloped regions.

2.5.7.6 Globalisation and De-globalisation Trends

Given shifting global trade dynamics, future research may explore how Asia's experience with both globalisation and de-globalisation influences current economic policies. Comparative studies on supply chain localisation, trade diversification, and resource security could reveal sustainable strategies for balancing local and global dependencies.

Highlights

- The future of business is being shaped by the dynamic growth trajectories of Asia's Tigers and Cubs.
- Businesses and policy-makers must leverage innovative practices.
- Technological advancements and digitalisation are critical to economic growth.
- Businesses that prioritise customer engagement, data-driven strategies, and technological integration will thrive.
- Sustainable growth relies not only on technological progress but also on cross-sectoral partnerships.

References

Antonio, R. (2021). Ligo sardines shows up what decisive youth leadership looks like in big business. *Scout Magazine*. https://www.scoutmag.ph/57067/ligo-sardines-show-us-what-decisive-youth-leadership-looks-like-in-big-business-1/

Asian Development Bank. (2020). *Asian Development Outlook 2020 update: Wellness in worrying times*.

Asian Development Bank. (2021). *Asian Economic integration report 2021: Making digital platforms work for Asia and the Pacific*.

Azoth Analytics (2023). *Global Artificial Intelligence Server Market (2023 Edition)*. https://www.marketresearch.com/Azoth-Analytics-v4068/Global-Artificial-Intelligence-Server-Edition-35657069/

Brimble, P., Jones, H., Don, U., & Zhang, X. (2022). Revisiting Thailand's economic development. *Economic Analysis and Policy, 73*, 320–336.

Business Sweden (n.d.). *The rise of the Southeast Asian Tigers* (2nd ed). https://www.business-sweden.com/contentassets/c5d9f4d114f14219a3f0be9c3ac80145/the-rise-of-the-southeast-asian-tigers.pdf

Camba, A. (2022). Platform governance in Southeast Asia: Data privacy, disinformation, and democracy. *Regulation & Governance, 16*(3), 847–864.

Chang, C., & Chen, R. (2022). Driving business model innovation of green product-service systems toward sustainability in Asia. *Sustainability, 14*(5), 2913.

Chang, H. J. (2022). The East Asian developmental state: The history and logic of the model. In *The political economy of capitalism: The Palgrave handbook* (pp. 577–597). Palgrave Macmillan.

Chaudhari, J. (2023). Industry 6.0: The rise of intelligent manufacturing and the future of industry. Retrieved April 6, 2025. https://medium.com/illumination/industry-6-0-the-rise-of-intelligent-manufacturing-and-the-future-of-industry-edb5f77f1706

Chew, J., Kittiprapas, C., & Xu, Q. (2022). Supply chain finance ecosystem resilience in Singapore. *Journal of Risk Finance, 23*(1), 60–79.

Cirineo, J. (2020). Brands & business: Ligo philippines redirects advertising funds to COVID-19 relief and response efforts. *Adobo Magazine*.

Corporate Finance Institute (CFI) (n.d.). Four Asian tigers. https://corporatefinanceinstitute.com/resources/economics/four-asian-tigers/

Dalton, M. (2022). How GoTo became Indonesia's largest tech company | Innovations case narrative. *Innovation Policy Platform*. https://www.innovationpolicyplatform.org/

Deloitte. (2022). *The Deloitte global 2022 Gen Z and millennial survey*. https://www.deloitte.com/global/en/pages/about-deloitte/articles/millennialsurvey.html

Dow Jones Institutional News. (2024, February 13). The Tiger cub economy hustling for its big break without Apple, Tesla—WSJ. *Dow Jones Institutional News*. https://www.proquest.com/wire-feeds/tiger-cub-economy-hustling-big-break-without/docview/2926038234/se-2.

Eggleston, K. (2023). Evidence-based public–private collaboration in the health sector: The potential for collaborative governance to contribute to economic recovery from COVID-19 in Asia. *Asian Development Bank South Asia Working Paper Series*. https://www.adb.org/sites/default/files/publication/861401/sawp-097-public-private-collaboration-health-sector.pdf. https://doi.org/10.22617/WPS230027-2

Endo, K., & Ram, K. S. (2021). Public–private partnerships in developing Asian countries: practical suggestions for future development assistance. Asian Development Bank Institute. https://www.adb.org/sites/default/files/publication/734081/adbi-brief-ppp-developing-asian-countries-3rd.pdf

Jun, S., & Suh, J. (2013). Exploring timing and environmental factors in new product adoption: The case of mobile TV. *Journal of Engineering and Technology Management, 30*(1), 72–89.

Karuppiah, K., Sankaranarayanan, B., & Ali, S. M. (2023). A systematic review of sustainable business models: Opportunities, challenges, and future research directions. *Decision Analytics Journal, 8*, Article 100272. https://doi.org/10.1016/j.dajour.2023.100272

Kimura, F., & Chen, L. (2018). Value chain connectivity in Indonesia: The evolution of unbundlings. *Bulletin of Indonesian Economic Studies, 54*(2), 165–192. https://doi.org/10.1080/00074918.2018.1505412

Liang, A. (2024, January 15). AI to hit 40% of jobs and worsen inequality, IMF says. https://www.bbc.com/news/business-67977967

Mahbubani, K. (2022). *Has the west led the world astray?* Polity Press.

Malesky, E. (2022). *The Vietnam development story: Economic reform as political learning.* Cambridge University Press.

Mandzieva, T., Tri, T., Kim, C., & Patel, R. (2018). Growth and integration of Southeast Asian economies. *Journal of Developmental Economics, 19*(3), 45–58.

Moher, D., Liberati, A., Tetzlaff, J., Altman, D. & for the PRISMA (2009). Group Preferred reporting items for systematic reviews and meta-analyses: the PRISMA statement. BMJ 2009; 339 :b2535. https://doi.org/10.1136/bmj.b2535

Nomura. (2024). Asia macro outlook 2025: Navigating the rapids. https://www.nomuraconnects.com/focused-thinking-posts/asia-macro-outlook-2025-navigating-the-rapids/

Ohno, K. (2009). Avoiding the middle-income trap: Renovating industrial policy formulation in Vietnam. *ASEAN Economic Bulletin, 26*(1), 25–43.

Paul, J., Lim, W., O'Cass, A., Hao, A., & Bresciani, S. (2021). Scientific procedures and rationales for systematic literature reviews (SPAR-4-SLR). *International Journal of Consumer Studies, 45*(4), O1–O16.

Peterson, T., Reyes, J. M., & Karoui, R. (2022). Corporate foresight and innovation of Asian battery companies: Lessons from LG Chem, Samsung SDI, and BYD auto. *Technological Forecasting and Social Change, 180*, Article 121719.

PR Newswire. (2016). Global growth opportunity: The rise of mobile & m-commerce in the Southeast Asian tiger cub economies. *PR Newswire*, April 19. https://www.proquest.com/wire-feeds/global-growth-opportunity-rise-mobile-amp-m/docview/1781812190/se-2

PwC. (2022). *The world in 2050.* https://www.pwc.com/gx/en/issues/economy/the-world-in-2050.html

Rasiah, R., Xiao, S., & Vonsiriporn, P. (2022). Long term view of Malaysia's catching up industrial policies. *Institutions and Economies*, 159–194.

Restrepo-Echavarría, P. & Arias, A. (2017) Tigers, Tiger Cubs and Economic Growth. *St. Louis Fed On the Economy.*

Rohrbeck, R., & Kum, M. (2018). Corporate foresight and its impact on firm performance: A longitudinal analysis. *Technological Forecasting and Social Change, 129*, 105–116.

Rohrbeck, R., & Lee, S. (2022). Corporate foresight and social responsibility: The case of the textile, apparel and luxury goods industry. *Technological Forecasting and Social Change, 177*, Article 121623.

Rui, Y., & Yao, J. (2016). Dragon and the tiger cubs. In *The handbook of global education policy* (pp. 385–400). John Wiley & Sons, Ltd.

Sachs, J. D., Schmidt-Traub, G., Fröhlich, C. K., & Krchniva, K. (2022). *The decade of action for the sustainable development goals.* Cambridge University Press.

Santos, F., & Yamada, K. (2023). COVID-19 impacts and building new resilience: A comparative analysis between Japan and Philippines. *Journal for ReAttach Therapy and Developmental Diversities, 6*(4s), 252–259. https://jrtdd.com/index.php/journal/article/view/441

Statista. (2023). Global E-Commerce Growth and Internet Usage. https://www.statista.com

Strange, R. (2020). The 2020 COVID-19 pandemic and global value chains. *Journal of Industrial and Business Economics, 47*(3), 455–465.

Stiglitz, J. E. (2022). *Globalisation and its discontents revisited: Anti-globalisation in the era of Trump.* W.W. Norton & Company.

Studwell, J. (2020). The institutional backdrop to Asian industrialisation. *Asian Economic Policy Review, 15*(1), 38–56.

Thuzar, M. (2021). Greening finance in Southeast Asia. *Asian Development Bank Institute Working Paper Series, 1159*, 1–19.

World Intellectual Property Organisation. (2022). *Global innovation index 2022: Who will finance innovation?* WIPO. https://www.wipo.int/global_innovation_index/en/2022/
World Wildlife Fund. (2020). In M. Grooten, & R. E. A. Almond (Eds.) *Living planet report—2020: Bending the curve of biodiversity loss.* WWF.
World Bank. (2024, October). Jobs and technology. The World Bank East Asia and the Pacific Economic Update. https://openknowledge.worldbank.org/server/api/core/bitstreams/b881d2ff-9912-4eb6-9698-8f151975abb6/content
Xiaolan, F., Zhang, J., & Chesbrough, H. (2021). China as a global Tech innovation Hub. *Research Policy, 50*(7), Article 104258.
Yusuf, S. (2022). Innovation systems and capabilities in developing East Asia. *Asian Development Review, 39*(1), 1–32.
Zhang, D., Si, L., Jie, J., & Yuhong, L. (2022). What accounts for regional resilience during the COVID-19 pandemic? A complex adaptive system perspective. *Sustainable Cities and Society, 77*, Article 103613.

Luz Suplico-Jeong has co-authored books on the Principles of Marketing and International Marketing. Not only is she a book author but she is a dedicated researcher on topics such as consumer behavior and export marketing. The research paper "Does Altruism Affect the Purchase Intent of Green Products", which she co-authored, was awarded "Outstanding Paper" by the National Academy of Science and Technology (NAST). NAST is the highest award-giving body for research in the Philippines. Presently, Dr. Suplico Jeong teaches marketing at De La Salle University (DLSU)-Manila. She is the Director of DLSU Center for Business Research and Development (CBRD).

Nelson B. Guillen Jr. is a certified marketing professional, educator, and researcher at De La Salle University, where he pioneered the Marketing and Advertising Research Sub-Center. He serves as CMO at NARRA Corporation and holds prestigious academic chairs. As Editor-in-Chief of three journals and Head of Student Engagement for PRME ASEAN+, Dr. Guillen contributes significantly to academia. His research focuses on social media marketing, brand management, and consumer behavior. With a Ph.D. in Business, M.S. in Marketing, and an M.A. in Digital Marketing, he continues to advance his marketing knowledge through his roles as a paper reviewer and active member of professional organisations.

Francia R. Santos is a Certified Professional Marketer Philippines. She holds a Master of Arts degree in Education major in Educational Psychology (UP-Diliman) and a Doctor of Business Administration (De La Salle University-Manila). She is a recipient of both Gold and Platinum Awards from the Global Sustainable Futures Network (GSFN 2021–2022) for her meaningful contribution as country coordinator for sustainability-related events of the global network. At present, she is an Associate Professor in the Department of Marketing and Advertising, Ramon V. del Rosario College of Business (RVRCOB) DLSU-Manila. She is currently the President of the Executive Council of Deans and Educators in Business (ECODEB) in the Philippines.

Chapter 3
Leadership Catalyst: Igniting Innovation in the Business Landscape

Clusters of Future Studies: Corporate Foresight; Past and Futures; Environmental Futures; Technological Trends

Syriac Nellikunnel Devasia and Soney Haris

Abstract In the contemporary business landscape, marked by constant change and heightened competition, this chapter conceptually and critically examines the indispensable role of leadership in steering organisations toward innovation and sustained success. The narrative draws on an extensive array of meticulously researched case studies and practical examples to elucidate the interplay between leadership prowess and innovation. It scrutinises how adept leaders inspire teams, cultivate an environment conducive to experimentation, and navigate challenges to foster a culture of continuous innovation. An additional focal point is the pivotal role of inclusive leadership in fostering innovation. By championing diversity, equity and inclusion initiatives; leaders can unlock a spectrum of perspectives and talent, fostering an environment where unique ideas flourish and thrive. In conclusion, "Leadership Catalyst: Igniting Innovation in the Asian Business Landscape" offers indispensable insights for leaders, executives, and aspiring change-makers navigating the complexities of the modern business terrain. This chapter serves as a strategic guide for unlocking the full potential of individuals, teams, and organisations in the relentless pursuit of sustainable and transformative innovation.

Keywords Leadership catalyst · Transformative innovation · Asian business · Cultural transformation · Organisational sustainability · Inclusion

S. N. Devasia (✉)
Perdana University, Kuala Lumpur, Malaysia
e-mail: syriac@perdanauniversity.edu.my

S. Haris
St. Joseph's College of Commerce, Bengaluru City University, Bengaluru, India
e-mail: drsoneyharis@sjcc.edu.in

M. S. Balaji et al. (eds.), *Future of Business in Asia*,
https://doi.org/10.1007/978-3-031-95940-0_3

3.1 Introduction

The dynamic landscape of business relies heavily on leadership as the engine of innovation. Leaders must become innovation catalysts as their organisations face previously unheard-of opportunities and challenges. The goal of this conceptual piece is to dissect the many facets of leadership that spark creativity and define the features of the corporate environment particularly in Asia.

3.2 Literature Review

3.2.1 The Leadership-Innovation Nexus

In today's quickly changing business environment, a dynamic nexus that affects an organisation's fate rests at the crossroads of leadership and innovation. The catalyst, or the power behind innovation that takes it from mere inspiration to implementable plans and revolutionary results, is leadership. Among the visionary leaders in this nexus are those who cannot only predict future trends but also effectively communicate a compelling vision that inspires their teams to pursue novel ideas. They see innovation as a strategic imperative which acknowledges that innovative solutions that redefine industries and generate new value propositions are necessary to stay ahead of the competition rather than just making small, incremental improvements.

The idea of visionary leadership emerges as a key component of the innovation-leadership relationship and explores the link between visionary leadership and employee creativity (Wang et al., 2022). Leaders with vision surpass the current state of affairs, seeing opportunities that others might miss. They present a future state that is in line with the demands of a changing society, growing market trends, and technology breakthroughs. This forward-thinking viewpoint serves as the organisation's compass amid ambiguity, fostering a culture of constant adaptation and development. In this setting, innovation gets embedded in the organisational DNA and permeates every facet of operations, rather than being limited to a single event or department.

One of the most important aspects of the relationship between leadership and creativity is team empowerment and shared vision (Van Dierdonck et al., 2020). Leaders who understand that innovation is a team effort to create a culture in which each member of the team feels both obliged and inspired to share their special knowledge. Leaders who empower their teams capitalise on the variety of skills present in their workforce that respect and celebrate diversity and creativity. Within the leadership-innovation nexus, this cooperative approach turns obstacles into opportunities, failures into teaching moments, and ignites the creativity needed for long-term organisational success.

3.2.2 *Defining Leadership Catalysts*

Organisational innovation is propelled by visionary individuals known as leadership catalysts, who go beyond conventional managerial responsibilities. These leaders have a special combination of qualities that push their people over the edge and create an atmosphere where new ideas germinate and creativity blossoms. Their unrelenting dedication to a compelling vision that inspires and motivates their teams toward a common goal—while also anticipating industry trends—is the foundation of their brand and nurture future-oriented leadership (DeRue et al., 2021).

Catalysts for leadership have a deep awareness of the value of innovation in navigating the intricacies of today's corporate environment. They understand that being a catalyst is more than just navigating the vessel; it also entails inspiring people to be creative and fostering an environment that encourages experimentation, accepts ambiguity, and learns from both achievements and setbacks. Thus, ethical leadership acts as a catalyst for employee creativity (Gong et al., 2020).

The key characteristic of leadership catalysts is their ability to amplify and empower their colleagues' potential. They go beyond conventional hierarchical systems and encourage team members to take responsibility for their actions. Leadership catalysts use their environment to bring out the best in their teams by actively seeking out and welcoming different points of view for leadership as a catalyst for individual learning agility (Afsar et al., 2020). This strategy changes company culture from the inside out by igniting creativity and fostering a feeling of purpose among everybody involved. To put it simply, the definition of a leadership catalyst is someone who creates change by navigating current obstacles and laying the foundation for a time when innovation will always be the driving force behind organizational success.

3.2.3 *Characteristics and Traits of Leaders Who Catalyse Innovation*

In the constantly changing world of business, leaders who function as catalysts for innovation possess a unique mix of qualities and attributes.

Curiosity and Lifelong Learning: Innovative catalysts are naturally curious people who are constantly looking to learn about new concepts, advancements in technology, and market trends. They promote an environment of ongoing education among team members and cultivate a culture of curiosity within the company.

Strategic Thinking: Strategic thinkers that are adept at connecting innovative projects with overarching organizational objectives are innovation catalysts. They strategically position their organisations for long-term success by anticipating possible future possibilities and obstacles.

Adaptive Leadership Style: Innovative leaders modify their style of leadership according to the scenario, understanding that certain circumstances could call for

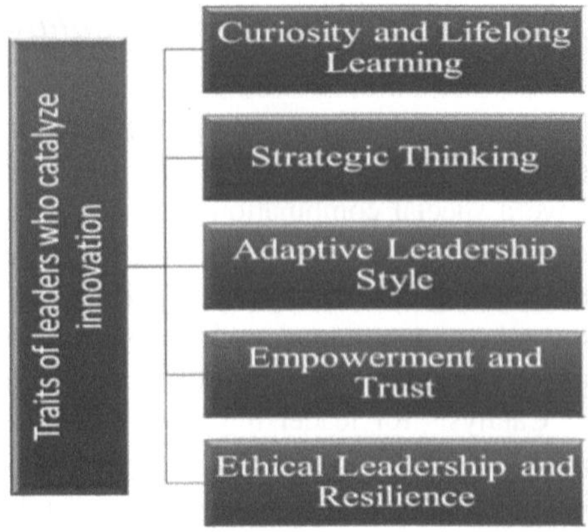

Fig. 3.1 Traits of leaders who catalyse innovation. *Source* Contributed by authors

different strategies. Depending on the needs of the innovation process, they can strike a balance between providing more facilitative leadership and providing hands-on coaching for cultivating adaptive expertise in leaders (Davis et al., 2021).

Empowerment and Trust: Innovation catalysts give their teams autonomy so they may make quick and right decisions. They foster an environment of trust where team members are free to offer ideas without worrying about receiving negative feedback.

Ethical Leadership and Resilience: Innovation catalysts base their decisions on ethical grounds. They place a high priority on honesty and moral behavior, making sure that innovation complies with social standards and the organisation's core principles. Innovation leaders are aware that pursuing novel ideas is inevitably fraught with setbacks. Because of their resilience, they are able to lead their teams toward future achievement, overcome obstacles, and learn from mistakes (Fig. 3.1).

In the Asian business context, the regional dynamics and diverse economic landscape has caught the attention on the role of leadership as a catalyst for innovation. Scholars and practitioners have highlighted the synergy of interplay among leadership styles, cultural factors and organisational capabilities in driving innovation insights, based on quickly evolving Asian markets.

3.2.4 Innovation and Leadership Styles

Bass and Riggio (2006) states that transformational leaders encourage employees to think creatively and embrace change and are shown by the leaders who practice them, such as, vision, inspiration, and intellectual stimulation. This approach is often adapted in the Asian context to reflect collectivist values; such as, teamwork, harmony (House et al., 2004). According to Li et al. (2018), leaders from China and Japan usually employ an authoritative and a participative style of leadership to act as an innovative force while at the same time bridging organisational coherence.

3.2.5 Influence of Cultural on Leadership and Innovation

Asia's cultural diversity presents unique challenges and opportunities for innovation. Leadership practices are dependent upon Hofstede's (1980) cultural dimensions, like power distance and collectivism. Leaders with cultural intelligence lead more effectively in their handling of different cultural environments and can maximise cross border collaboration and innovation (Earley & Ang, 2003). For example, leaders in Southeast Asia often leverage cultural nuances, such as respect for hierarchy and community orientation, to build cohesive teams that can innovate effectively (Nguyen et al., 2020).

3.2.6 Technological Advancements and Leadership Adaptation

In Asia, the rapid adoption of digital technologies has transformed the role of leaders into those, who drive innovation. With artificial intelligence, blockchain, and big data, the digital age has introduced new disruptive technologies like leadership in the digital age would need to be adaptable, should have strategic foresight, and should be willing—and should be ready, to some degree—to adopt artificial intelligence, blockchain, and big data (Chuang et al., 2021). To sustain their competitive advantages and push the boundary of innovation, companies such as Huawei and Tencent gives a perfect example of how leaders can exploit digital transformation (Zhang & Wang, 2020).

3.2.7 The Innovation Imperative

The rate of change confronting organisations is always increasing due to factors such as global interconnection, evolving consumer expectations, and technology breakthroughs. In light of this, the innovation imperative ceases to be a discretionary endeavor and instead becomes a strategic necessity. These days, being able to innovate is more than just a mark of industry leaders—it's what keeps companies relevant and resilient in a disruptive age. Innovation also calls for taking calculated risks. For example, the founder and CEO of SoftBank Group, Masayoshi Son, is renowned for taking a risk-taking and forward-thinking strategy when making investments in technology startups. Under his direction, SoftBank has grown to be a significant force in the global tech sector, investing in innovative businesses such as ARM Holdings, Uber, and Alibaba. Son is well-known in the innovation industry thanks to his willingness to take chances and his excellent sense of disruptive technology (Bloomberg, 2024).

The realisation that falling behind is the same as remaining still is at the core of the innovation imperative. With the increasing obsolescence of established business models and the development of new, disruptive players, industries are undergoing fundamental upheavals. To put it simply, the innovation imperative is a call to action that challenges companies to change, adapt, and reinvent themselves in order to satisfy the ever-changing needs of the contemporary marketplace. For example, Tencent Holdings CEO and co-founder Pony Ma has been instrumental in forming China's tech sector. Some of the most widely used internet services in China, such as WeChat, QQ, and Tencent Games, are owned by Tencent. Tencent has become a global leader in social networking, gaming, and digital entertainment under Ma's direction, reaching beyond China. Ma's astute insight and well-timed investments have established Tencent as a major participant in the international digital economy (Ma Huateng, 2024).

Businesses that do not adopt innovation run the risk of becoming obsolete and stagnant, and they will be surpassed by rivals that recognise innovation as a necessity rather than an extravagance. The innovation imperative, therefore, cuts across industry lines, requiring firms to foster a culture that appreciates creativity, welcomes change, and actively looks for chances for expansion and improvement while discussing the challenges and opportunities in measuring the social impacts of innovative sustainability practices (Epstein & Buono, 2020). For example, Reliance Industries' founder, Dhirubhai Ambani, is recognised as one of India's most creative businessmen. Ambani turned Reliance from a modest trading business into a conglomerate with businesses in retail, telecommunications, refining, and petrochemicals. A new generation of entrepreneurs has been greatly inspired by his entrepreneurial drive and capacity to upend well-established sectors, which has had a significant impact on India's commercial environment (Dhirubhai Ambani Biography, Accomplishments, and Facts, 2024).

3.2.8 *Reflection: Why is Innovation Crucial in the Contemporary Business Landscape?*

Innovation is essential to survival and success in the fast-paced, intricate business environment of today. It is not just a competitive advantage. A number of strong arguments highlight how important innovation is in determining the future course of businesses and whole sectors.

Rapid Technological Advancements: One of the things that sets the present era apart is the unrelenting speed at which technology is developing. Businesses that don't innovate run the risk of being rendered obsolete by technology and losing out on opportunities to benefit from innovations that improve productivity, customer satisfaction, and overall competitiveness.

Changing Customer Expectations: Modern customers have higher standards since they have access to a multitude of options and are empowered by digital connectedness. Meeting and surpassing these expectations require innovation, whether it be in the form of introducing new products and services, delivering individualised experiences, or facilitating seamless digital interactions. Businesses that innovate not only win over new customers, but also establish service-dominant logic as a framework for understanding changing expectations (Pentina & Taylor, 2020).

Agile Response to Uncertainty: The acronym VUCA stands for volatility, uncertainty, complexity, and ambiguity, which characterises the modern business environment. Static business models don't work in such a setting. Innovation allows firms to react quickly to changing conditions because it emphasises flexibility and agility. Instead of only responding to outside influences, companies that cultivate an innovative culture are better equipped to handle uncertainty, change course when needed, and take control of their own destiny.

Evolving Regulatory and Compliance Landscape: Global industries are impacted by the constant evolution of regulatory requirements and compliance standards. In order for firms to meet regulatory requirements and take advantage of compliance as a chance to stand out from the competition, innovation is essential. Companies that innovate in their compliance procedures can improve operations, increase transparency, and create a solid foundation that can survive changes in regulations.

Talent Attraction and Retention: Top talent in today's workforce looks for conditions that encourage innovation, education, and personal development. Talented workers wanting to contribute to ground-breaking ideas are drawn to innovative firms. Businesses that prioritise innovation attract top talent and foster a culture that maintains and develops the most brilliant brains.

Thus, innovation is critical to an organisation's ability to respond to unpredictability, satisfy changing customer expectations, traverse global complications, respond to technological improvements, adjust to regulatory changes, and recruit top talents. This makes it a strategic imperative in today's business world. In a world where change is relentless and innovation is correlated with resilience and long-term success, organisations that embrace and value innovation not only survive but flourish.

3.3 Methodology

A case study approach is taken to shed light on the role of leadership in steering organisations toward innovation and sustained success. Cases are selected based on companies operating in Asia across a wide range of industries. Cases are written in a structured and reflective manner, where a brief background is provided for each case in point and then an expert commentary based on the authors' input is provided.

3.4 Findings

The following findings from the case studies are presented through the lenses of innovation, competitiveness, and organisational sustainability, as well as the roles of visionary and transformative leadership, including the influence of leadership catalysts as these dimensions provide a comprehensive framework to understand how strategic actions drive impactful and enduring organisational outcomes.

3.4.1 Innovation, Competitiveness, and Organisational Sustainability

3.4.1.1 Samsung Electronics (South Korea)—Consumer Electronics and Semiconductors

Samsung Electronics is a global leader in consumer electronics, mobile devices, and semiconductor technology. Global consumer preferences and industry norms have been formed by the company's inventive goods, which include memory chips, televisions, and cellphones. Samsung's competitive advantage in the ever evolving technological sector stems from its dedication to innovation and emphasis on research and development (Abraham et al., n.d.).

3.4.1.2 Infosys (India)—Information Technology and Consulting Services

Renowned for its inventive approach to digital transformation and business process outsourcing, Infosys is a leading global player in information technology and consulting services. Utilising cutting-edge technology like blockchain, cloud computing, and artificial intelligence, the business assists customers in streamlining their operations and spurring expansion. As one of the most admired organisations in the world of IT services, Infosys has gained reputation for its dedication to innovation and excellence (Infosys—Consulting, IT Services, Digital Transformation, n.d.).

3.4.1.3 Reflections

The foundation of long-term success in the complex web of the modern business environment is the relationship between innovation, competitiveness, and organisational sustainability. This mutually beneficial relationship is both strategic and intuitive, as businesses understand that their capacity for innovation is a key factor in determining both their long-term survival and competitive position.

Enhancing Competitiveness through Differentiation: Organisations can effectively differentiate themselves in a crowded and competitive environment by utilising innovation as a potent strategy. By launching innovative goods, services, or methods, companies distinguish themselves from competitors. In addition to providing a competitive edge, this distinction serves as a spur for gaining market share and building client loyalty and understanding performance under varying market conditions (Chen et al., 2021).

Optimising Operational Efficiency: Innovation includes streamlining internal procedures and workflows in addition to providing new or improved products and services. Innovative operational solutions improve productivity, cut expenses, and simplify processes for organisations. By increasing efficiency, this operational excellence not only boosts competitiveness but also lays the groundwork for long-term sustainability.

Strategic Positioning in Global Markets: Organisations operate in a borderless environment in the age of globalisation, where rivals can appear from anywhere in the world. When it comes to placing companies strategically in international marketplaces, innovation is essential. Innovative businesses go beyond national borders to fortify their competitive positions, whether it is by developing internationally scalable technologies or customising products for varied cultural contexts.

Building Resilience for Long-Term Sustainability: The resilience that creative practices foster is the foundation for the relationship between innovation and organisational sustainability. Companies that innovate constantly are able to withstand unforeseen difficulties, economic ups and downs, and upheavals. Organisations that are able to innovate and adapt will not only survive but also grow throughout by measuring the social impacts of innovative sustainability practices (Epstein & Buono, 2020).

Strategies for Leadership Catalysts: A variety of strategies are used by leadership catalysts to promote innovation in their companies. A strategic alignment that directs team efforts toward transformative objectives is produced by leadership catalysts by cultivating a common sense of the organisation's future course. This strategic clarity gives teams a sense of purpose and motivation to overcome obstacles and take advantage of opportunities, which eventually propels the firm into new areas of innovation.

A fundamental tactic for leadership accelerators is team empowerment. Understanding that innovation is a team effort, effective leaders foster a culture in which team members feel both empowered and obligated to share their perspectives. Through the decentralisation of decision-making and the promotion of autonomy, leaders may effectively leverage the different skills present in their teams. This method not only encourages innovation but also cultivates a culture in which each person is valued for their capacity to provide novel solutions.

Another important tactic used by leadership catalysts to negotiate the intricacies of the innovation environment is agile leadership. Agility is critical in a company environment that is changing quickly. Catalysts for leadership adopt a flexible mindset that allows them to quickly adjust to changing conditions and try to measure how psychological capital acts as a catalyst for transformational leadership (Dinh et al., 2021). They provide adaptability in procedures and organisational structures so that groups can try new things, iterate, and gain knowledge from both successes and setbacks.

3.4.2 Visionary Leadership

3.4.2.1 BYD Company Limited (China)—Electric Vehicles and Renewable Energy

One of the top producers of renewable energy sources and electric cars is BYD Company Limited. The company's cutting-edge electric vehicles, which include trucks, buses, and passenger cars, are becoming more and more well-liked both domestically and abroad. By producing batteries and other essential components internally, BYD's vertically integrated strategy has allowed it to cut costs and spur innovation in the electric vehicle market (Pereira, 2024).

3.4.2.2 Toyota Motor Corporation (Japan)—Hybrid and Electric Vehicles

Toyota Motor Corporation has led the way in innovation in the automotive sector when it comes to the creation of hybrid and electric cars. The company's ground-breaking Prius hybrid vehicle, which debuted in the late 1990s, cleared the path for hybrid technology to become widely used in the automobile industry. Toyota is committed to advancing electric vehicle technology via research and development, with the aim of mitigating carbon emissions and advancing sustainable transportation alternatives (Ashley Lobo, 2020).

3.4.2.3 Reflections

The transformative potential of visionary leadership is at the core of the techniques employed by leadership catalysts. Leaders with a clear and inspirational vision steer their organisations through unfamiliar territory, creating the future. This kind of leadership goes beyond the ordinary and customary, driving groups of people toward a common goal that goes beyond the problems at hand. Leaders with a clear vision know that innovation requires more than just small adjustments; it also requires a deep, long-term view that accounts for changes in the market and new trends.

The capacity to passionately and clearly convey a compelling vision is a distinguishing feature of visionary leadership. These executives provide a compelling picture of the organisation's goals that goes beyond the details of day-to-day operations. Visionary leaders generate a feeling of purpose that permeates the entire organisation by cultivating a common understanding of the destination. This unambiguous vision serves as a focal point, uniting disparate teams behind a shared objective and igniting a passion for innovation among all of them. Organisations with visionary leadership aggressively seek out chances for growth and advancement and accept change as a necessary part of life, paving the way for long-term success and sustainable innovation.

In order to foster innovation in organisations, visionary leaders are essential. Their methodology entails a comprehensive examination and tactical execution to establish a setting that fosters innovative concepts and game-changing resolutions.

The cultivation of a culture that encourages creativity and taking risks is a critical component of how visionary leaders create the conditions for innovation. They are aware that an environment where workers are encouraged to experiment with new concepts without worrying about failing is conducive to innovation. Visionary leaders establish a psychological safety net that empowers teams to go beyond traditional bounds by deliberately fostering a culture that sees failures as teaching opportunities rather than roadblocks. This change in culture fosters experimentation, allowing people to explore the boundaries of their creativity and add to an ongoing cycle of invention.

Visionary leaders also make investments in building a solid infrastructure to foster innovation. This involves giving specialised innovation projects access to financial and human resources. Visionary leaders guarantee that there is a specific area for investigating and testing innovative concepts by establishing specialist teams or departments dedicated to research and development.

Moreover, visionary leaders support a strategic alignment that links innovation to the organisation's overall objectives. They clearly lay out their goal, emphasising how crucial innovation is to long-term success. This alignment makes sure that innovation is viewed as an essential component of organisational strategy rather than as a distinct, stand-alone role. A common goal that drives the entire organisation toward a future characterised by constant evolution and advancement is inspired by visionary leaders who effectively convey how innovation contributes to the company's growth and sustainability.

3.4.3 *Transformative Leadership*

3.4.3.1 Wan Zulkiflee Wan Ariffin—Petroliam Nasional Berhad (Petronas) (Malaysia)

With his progressive approach to the energy sector, Wan Zulkiflee Wan Ariffin, the former President and CEO of Petroliam Nasional Berhad (Petronas), epitomises transformative leadership. Petronas has successfully navigated obstacles in the global energy landscape under his leadership, positioning the company for innovation and sustainable growth. Noteworthy initiatives and innovations are as follows.

Energy Transition: Wan Zulkiflee saw the value of embracing the shift in Petronas' energy supply to cleaner, more sustainable sources and diversifying the company's energy portfolio. He led the charge to invest in green energy projects including wind and solar power in addition to Petronas' conventional oil and gas business. These investments support international initiatives to lower carbon emissions and slow down climate change.

Digital Transformation: Wan Zulkiflee gave digital transformation top priority at Petronas, using data analytics and technology to improve productivity, spur innovation, and optimise operations. Petronas initiated many projects, including the creation of digital platforms and solutions, with the aim of optimising operations, enhancing decision-making, and generating value throughout the whole value chain.

Involvement of Stakeholders: Wan Zulkiflee stressed the value of cooperation and involvement with stakeholders in Petronas' operations. In order to advance shared value creation, social responsibility, and sustainable growth, he promoted collaborations with local communities, industry stakeholders, and governmental organisations. Wan Zulkiflee's dedication to ethical business practices and corporate responsibility is reflected in Petronas's programs for environmental conservation, education, and community development.

The leadership exhibited by Wan Zulkiflee Wan Ariffin at Petronas highlights the significance of innovation, strategic planning, and forward-thinking vision in promoting organisational success and long-term growth in the energy industry. Because of his efforts, Petronas is now a major player in the global energy transition, proving the transformative power of visionary leadership in overcoming obstacles in the business and grasping chances for future success (Petroliam Nasional Berhad (Petronas), 2023).

3.4.3.2 Masayoshi Son—SoftBank Group (Japan)

SoftBank Group's founder and CEO, Masayoshi Son, is renowned for his audacious and innovative management style, which has helped to build SoftBank into a major international technology corporation. Son has a forward-thinking strategy that includes investments in disruptive business models and transformative technologies that have altered industries and opened up new markets.

Initiatives and Innovations:

Vision Fund: Son established the SoftBank Vision Fund, a sizable investment vehicle devoted to providing funding to creative start-ups and cutting-edge technology businesses across the globe. Investments made by the Vision Fund in businesses like Uber, WeWork, and Slack have upended established markets and quickened the rate of technological advancement.

5G and IoT: Son has been an outspoken supporter of the creation and implementation of Internet of Things (IoT) and 5G networks. The goal of SoftBank's investments in

Table 3.1 Initiatives and innovations of transformative leadership through forward-looking vision

Organisation	Leadership	Initiatives and innovations
Petroliam Nasional Berhad (Petronas) (Malaysia)	Wan Zulkiflee Wan Ariffin	Energy transition
		Digital transformation
		Involvement of stakeholders
SoftBank Group (Japan)	Masayoshi Son	Vision fund
		5G and IoT
		Renewable energy

Source Contributed by authors

IoT platforms and telecommunications infrastructure is to build linked ecosystems that promote productivity, efficiency, and connectivity among various businesses.

Renewable Energy: SoftBank's investments in wind and solar power projects are one way that Son has shown his dedication to sustainability and renewable energy. Son wants to solve environmental issues and provide a more sustainable future for future generations by utilising renewable energy technologies.

These case studies show how visionary CEOs may revolutionise industries, promoting creativity and changing businesses. These CEOs not only accomplished great success for their firms through the development of a strong vision and the implementation of strategic initiatives, but they also left a lasting impact within their respective sectors. Their ability to foresee new trends, express a bold vision, and foster an inventive culture is a prime example of the transformative power of forward-thinking leadership. (Softbank Annual Report, 2024) (Table 3.1).

3.4.3.3 Reflections

Strategies for Leaders to Empower and Inspire Their Teams

Beyond administrative duties, effective leadership entails motivating and enabling groups to realise their greatest potential. Establishing a culture of open communication and attentive listening is one important tactic. Leaders who actively interact with their teams, soliciting feedback and appreciating other points of view, foster an atmosphere in which each team member feels appreciated and heard. Leaders establish trust by creating a psychologically secure environment where team members may voice concerns and ideas without fear of being judged. This allows individuals to bring their special talents to the table and contribute to the team.

Leaders may empower their teams not just by fostering an inclusive workplace but also by offering chances for skill development and ongoing education. This entails spending money on workshops, training courses, and other materials that improve team members' skills. People become more self-assured and motivated when they observe that their leaders are dedicated to their professional development and actively assist their learning processes. Teams that feel empowered are more likely to take responsibility for their job, view obstacles as chances for improvement, and be motivated to offer creative solutions that advance the organisation as a whole (Heathfield, 2021).

Clear Communication of Vision and Goals: A compelling vision and attainable goals should be clearly communicated by leaders to their teams. Team members are more likely to feel motivated and a sense of ownership when they are aware of the larger purpose and goals. A common goal fosters alignment and offers a path forward for group initiatives.

Delegate Authority and Foster Autonomy: Giving teams the power and responsibility to make decisions is a key component of empowerment. In order to give team members a sense of ownership over projects, leaders should assign duties according to their skills. This delegation fosters a culture of self-direction by encouraging a sense of accountability and autonomy in addition to boosting confidence (Deci & Ryan, 1987).

Recognition and Celebration of Achievements: Acknowledging and applauding accomplishments on an individual and team level is a strong source of motivation. Team members' efforts and hard work should be recognised and valued by leaders. Receiving praise from the public not only raises spirits but also emphasises how crucial each member is to the team's overall success.

Provide Opportunities for Learning and Development: Leader's ought to make an investment in their staff' ongoing education and growth. This can be mentorship opportunities, workshops, or training courses. Providing opportunities for skill development not only empowers individuals but also enhances the team's overall competency and adaptability (Carmeli et al., 2021).

Leaders may empower, inspire, and motivate their people to reach their maximum potential by putting these tactics into practice. A high-performing and engaged team is built on a foundation of clear communication, support, acknowledgment, and a dedication to both individual and group growth.

The Role of Agility in Leadership: Navigating Uncertainty and Fostering Adaptability

Agile leadership emphasises adaptability, collaboration, and iterative progress. It is a dynamic and responsive style that is in line with the concepts of agile methodology. Agile leadership has emerged as a critical way of thinking for guiding companies through complexity and uncertainty in the face of a business environment that is changing quickly. The ability to accept change as a given and to recognise that flexibility and adaptability are necessary traits for overcoming the obstacles of today's fast-paced and unpredictable surroundings is a crucial component of agile leadership.

Agile leaders prioritise collaboration and cross-functional teamwork. By breaking down silos and fostering a culture of open communication, agile leadership ensures that information flows seamlessly across organisational boundaries. This collaborative approach is instrumental in leveraging the diverse skills and perspectives of team members, promoting a collective sense of ownership, and enabling the or Cross-functional teamwork and collaboration are given top priority by agile leaders. Agile leadership makes sure that information moves freely across organisational boundaries by dismantling organisational barriers and promoting an environment of open communication (Rigby et al., 2020). Adopting the mindset of "fail fast, learn faster," agile leaders cultivate an environment that appreciates trial and error and sees setbacks as chances for growth rather than failures. In a constantly changing business environment, this iterative mindset helps firms stay flexible, react to new trends, and keep a competitive edge.

Navigating Uncertainty: In today's corporate world, uncertainty is a given, driven by things like new developments in technology, changes in the world economy, and unanticipated catastrophes like the most recent epidemic. Agile leadership recognises that the market is inherently unpredictable and provides organisations with the necessary resources to effectively manage uncertainty. Agile leaders promote a flexible mindset that enables prompt adjustments in reaction to changing circumstances, as opposed to rigorously adhering to preset plans.

Fostering Adaptability: Agile leadership places a strong focus on encouraging organisational agility. This entails fostering a culture that views change as a chance for development rather than as a danger. Agile leaders promote an approach that values ongoing experimentation and learning. Teams are given the freedom to modify their plans in response to immediate input, guaranteeing that the company will always be able to adjust to changing market conditions.

Enhancing Decision-Making Speed: One characteristic of agile leadership is its capacity for quick decision-making. Delays in decision-making can result in lost opportunities or inadequate solutions to problems in hectic work situations. Agile leaders minimise the need for hierarchical approvals by enabling teams to make choices at the right levels (Klein, 2023). Because of this expedited decision-making process, the company can move quickly and take advantage of possibilities before rivals do. The emphasis on speed is meant to foster a culture that prioritises agility and efficiency in the decision-making process rather than hastily making decisions.

Building Resilience: Effective leadership has a crucial role in fostering organisational resilience. Agile leaders perceive interruptions as chances to grow and adapt, not as insurmountable impediments. Being resilient involves learning how to overcome adversity and come out stronger rather than trying to avoid problems. Agile leaders make sure that teams can weather storms, learn from experiences, and emerge more adaptive and prepared for future uncertainty by fostering a resilient attitude inside the firm.

3.4.4 *Leaders Who Successfully Shifted Their Organisational Culture Towards Innovation*

3.4.4.1 Ren Zhengfei—Huawei Technologies (China)

The founder and CEO of Huawei Technologies, Ren Zhengfei, has changed the company's organisational culture to emphasise creativity and adaptability. Ren places a strong emphasis on long-term planning, funding R&D, and cultivating an environment that is open to ongoing learning and change. Noteworthily initiatives and innovations are as follows.

Investment in Research and Development: Ren Zhengfei has made research and development a top priority, dedicating large sums of money to spur the development of cutting-edge telecoms technologies like 5G, AI, and cloud computing. Huawei's commitment to research and development has allowed it to create innovative goods and services that can compete on a global scale.

Entrepreneurial Culture: Ren has cultivated an entrepreneurial culture at Huawei, giving staff members the freedom to take initiative, follow their ideas, and question the established status quo. Because of Huawei's meritocracy and flat organisational structure, workers have many chances to contribute their knowledge and skills to the company's development.

International Cooperation: Ren Zhengfei has underlined the value of international cooperation and collaborations with clients, vendors, and business partners. Huawei has accelerated innovation and kept the corporation at the forefront of technical improvement through its research, standardisation, and ecosystem development collaborations with worldwide partners.

Outcomes: With a presence in more than 170 nations and regions, Huawei has emerged as a global leader in telecommunications technology under the direction of Ren Zhengfei. Huawei has established itself as a major participant in the digital transformation age with its cutting-edge technologies and solutions, worldwide collaboration, and dedication to customer-centricity (https://www.huawei.com/en/annual-report/2023).

3.4.4.2 Ciputra—Ciputra Group (Indonesia)

Ciputra is a well-known businessman from Indonesia who founded the Ciputra Group, a varied conglomerate with holdings in real estate, hotels, property development, and education. Ciputra has shown visionary leadership and a dedication to innovation throughout his career, influencing the Ciputra Group's organisational culture to value innovation, entrepreneurship, and sustainable development. Noteworthily initiatives and innovations are as follows (CTRA Annual-Report 2023).

Entrepreneurial environment: Ciputra has developed an innovative and risk-taking attitude among its business partners and workers by creating an entrepreneurial environment within the Ciputra Group. The group's many enterprises, which include real estate development, retail, and education, give ambitious business owners a platform to explore their concepts and endeavors, spurring growth and innovation.

Sustainable Development: Ciputra has been an advocate for environmental preservation and sustainable development, including sustainability ideas into the organisation's operations and commercial procedures. Ciputra is dedicated to generating long-term value with the least amount of negative environmental impact. This is shown in initiatives like green building design, the adoption of renewable energy, and community participation programs.

Investment in Education: Ciputra understands the value of education in promoting the growth of human capital and innovation. The Ciputra Group has founded educational establishments that prioritise entrepreneurship, critical thinking, and innovation, including schools and universities. These educational establishments equip students with the necessary information and abilities to prosper in a world that is changing quickly and to lead innovation in their industries.

Results: The Ciputra Group has grown to become one of Indonesia's top conglomerates, with a varied portfolio of businesses and a reputation for excellence and honesty, thanks to Ciputra's leadership and dedication to innovation. The group's creative endeavors and sustainable development programs have boosted the local economy, created jobs, and improved community well-being, illuminating the transformative potential of visionary leadership in fostering business success and societal impact.

The case study from Ciputra demonstrates how innovative thinking and visionary leadership may influence corporate culture and promote long-term, steady growth and development. Ciputra has left behind a legacy of innovation and excellence that continues to motivate future generations of business leaders and entrepreneurs in Indonesia and abroad. These initiatives include supporting sustainable development, encouraging an entrepreneurial ecosystem, and investing in education (Table 3.2).

Table 3.2 Leaders who successfully shifted their organisational culture towards innovation

Organisation	Leadership	Initiatives and innovations
Huawei Technologies (China)	Ren Zhengfei	Investment in research and development
		Entrepreneurial culture
		International cooperation
Ciputra Group (Indonesia)	Ciputra	Entrepreneurial environment
		Sustainable development
		Investment in education

Source Contributed by authors

3.4.5 *Leadership Catalysts*

3.4.5.1 Hiroshi Mikitani—Rakuten (Japan)

The founder and CEO of Rakuten, the biggest e-commerce business in Japan and a top provider of internet services globally, is Hiroshi Mikitani. Under Mikitani's direction, Rakuten became a worldwide digital ecosystem with a strong emphasis on innovation, internationalisation, and customer-centricity. Noteworthily initiatives and innovations are as follows (Global. Rakuten 2023).

Rakuten Ecosystem: Mikitani has developed Rakuten into an all-encompassing digital services ecosystem that encompasses finance, e-commerce, digital content, and telecoms. Customers can access a variety of goods and services because to Rakuten's platform strategy, which encourages interaction and loyalty.

Investment in Technology: Mikitani has made innovation and technology a top priority. To improve Rakuten's services and the consumer experience, the company is utilising machine learning, artificial intelligence, and data analytics. Research and development expenditures made by Rakuten have produced cutting-edge goods and services that set it apart from rivals.

Global Expansion: Mikitani has spearheaded Rakuten's efforts to take the digital economy worldwide by leading the company's push outside of Japan. Rakuten's expansion and competitiveness have been fueled by its collaborations and acquisitions with foreign businesses, which have allowed it to develop a presence in important global markets and diversify its revenue streams.

Results: Rakuten has expanded from a small online marketplace to a global digital powerhouse with businesses spanning e-commerce, finance, digital content, and telecommunications under Hiroshi Mikitani's direction. Rakuten is now positioned as a leader in the age of digital transformation because to Mikitani's innovative leadership and vision.

3.4.5.2 Kiran Mazumdar-Shaw—Biocon Limited (India)

Established in India, Biocon Limited is a leading biotechnology business that leads the world in biosimilars. Kiran Mazumdar-Shaw serves as its founder and executive chairperson. Biocon became a major player in the global biotech business under Mazumdar-Shaw's leadership because of his unwavering focus on innovation, entrepreneurship, and social impact. Noteworthily initiatives and innovations are as follows (Biocon Integrated Annual Report (2023–2024)).

Biosimilars and biopharmaceuticals: In order to fill unmet medical needs and increase access to reasonably priced healthcare, Mazumdar-Shaw has led the way in the development and marketing of biosimilars and biopharmaceuticals. Innovative products from Biocon, such as insulin, recombinant proteins, and monoclonal antibodies, have changed patient care and gained international attention.

Research and Development: Mazumdar-Shaw has made significant investments in this area, creating a top-notch innovation ecosystem that promotes the identification and creation of cutting-edge treatments. Modern R&D facilities at Biocon, along with partnerships with universities and research groups, have produced ground-breaking discoveries and biotechnology patents.

Social Responsibility: Through Biocon's programs in healthcare, education, and environmental protection, Mazumdar-Shaw has shown a dedication to sustainability and social responsibility. Biocon's patient support programs, inexpensive healthcare initiatives, and community involvement activities are indicative of Mazumdar-Shaw's philosophy for leveraging business to create positive social impact.

Under Kiran Mazumdar-Shaw's direction, Biocon has grown into a major player in the biotech industry with a portfolio of cutting-edge products and a solid reputation for honesty and quality. Mazumdar-Shaw's innovative spirit and imaginative leadership have made Biocon a global leader in the biotechnology sector and an advocate for accessible healthcare.

These case studies on Hiroshi Mikitani and Kiran Mazumdar-Shaw demonstrate how innovative thinking, smart investments, and visionary leadership can propel change and expansion in businesses. Mikitani and Mazumdar-Shaw have been key figures in influencing the direction of their respective businesses and bringing about constructive social change as catalysts for leadership.

3.5 Discussion of Implications

3.5.1 Overcoming Challenges to Innovation

Though it has its obstacles, innovation is a revolutionary force that drives enterprises ahead. Agile leaders are essential in helping to overcome these challenges because they cultivate an environment that values creativity and overcomes setbacks with fortitude.

Cultural Resistance to Change: Organisational resistance to change is a common cause of one of the main obstacles to innovation. Agile leaders actively create a culture that encourages experimentation and adaptation in order to address this. They spread the word about the value of innovation as a catalyst for long-term success and growth, encouraging people to view change as an essential and positive part of development rather than as a danger. Agile leaders dismantle resistance and foster an atmosphere where creativity may thrive by establishing a culture that embraces learning from mistakes and seeing setbacks as chances for development (Bunce & Roos, 2022).

Fig. 3.2 Various ways of overcoming challenges to innovation. *Source* Contributed by authors

Lack of Resources and Investment: Inadequate financial and human resources can hinder innovation efforts. To address this issue, agile leaders promote resource allocation that is strategic and gives priority to innovation. They strive to obtain the funds required for training initiatives, technological adoption, and research and development. Furthermore, agile executives promote cross-functional cooperation by utilising the organisation's current skills to spur innovation.

Resistance to Risk and Fear of Failure: Fear of failure can be a major impediment to creativity since it makes people and groups reluctant to take chances that can result in failure. In order to overcome this difficulty, agile leaders establish a psychological safety net for their team members. They stress that failure is a necessary component of the creative process and a chance to gain insightful knowledge. Agile leaders provide an example of a way of thinking that promotes experimentation and measured risk-taking. Agile leaders enable their teams to overcome risk aversion and welcome the uncertainty that accompanies innovation by creating a climate in which members feel supported and encouraged to experiment without worrying about negative consequences (Nixon & Noble, 2022) (Fig. 3.2).

3.5.2 Navigating Setbacks and Challenges in the Pursuit of Innovation

On the road to innovation, resilient leaders are essential in helping their people overcome obstacles and disappointments. Their capacity to persevere through hardship with grace and tenacity plays a critical role in creating an environment that values experimentation, education, and eventually ground-breaking innovation.

Embracing a Growth Mindset: Resilient leaders see obstacles as chances for growth and learning, and they address setbacks with a growth attitude. Failures are not seen as obstacles, but rather as insightful input that can guide future innovation initiatives. Resilient leaders inspire their team members to take measured risks and investigate novel ideas by cultivating a culture that values learning from failures. This growth attitude fosters a never-ending cycle of invention and progress in addition to enhancing resilience on both an individual and organisational level.

Promoting a Culture of Experimentation: Innovation by its very nature involves risk and uncertainty. Understanding this, resilient leaders aggressively encourage a culture of innovation among their staff members. They foster an environment that is risk-free for experimenting, taking initiative, and stimulating creativity. Resilient leaders help their teams learn from past mistakes by conducting in-depth post-mortems, drawing lessons, and using the information to improve future innovation tactics and see how leaders can create an environment conducive to experimentation amidst complexity (Heifetz & Linsky, 2020). Through this iterative approach, failures are turned into opportunities for successful innovation rather than insurmountable roadblocks.

Effective Communication and Transparency: When faced with obstacles, resilient leaders put good communication and transparency first. They communicate candidly with their colleagues about their accomplishments and mistakes, openly acknowledging obstacles. Open communication fosters resilience and trust inside the company, empowering people and fostering a culture of knowledge. Resilient leaders show vulnerability and authenticity by talking candidly about setbacks and the lessons they've learned.

Providing Support and Encouragement: Resilient leaders offer their teams unwavering support and encouragement when faced with obstacles. They make sure that team members feel supported both emotionally and professionally since they are aware of the emotional toll that obstacles may have on people. Even in the face of losses, resilient leaders celebrate accomplishments and efforts, fostering a supportive environment. In addition to boosting team spirit, this support fosters a resilient culture in which people are more inclined to overcome failures with renewed resolve.

Adapting Strategies and Iterating: Adaptability is exhibited by resilient leaders who are prepared to modify their plans in light of the lessons learned from failures. They provide direction to their teams as they iteratively improve and optimise innovation approaches. Resilient executives make sure that the company stays flexible and adaptable to changing conditions by altering tactics in reaction to obstacles. This flexible strategy helps the innovation process remain resilient overall by enabling the company to quickly bounce back from setbacks and move forward with more successful plans (Kumar & Sharma, 2023).

3.6 Future Directions

3.6.1 *Cultural Transformation*

An organisation's identity is defined by its values, beliefs, and practices, and cultural transformation is a significant change in these areas. The realisation that the current culture might not be in line with the organisation's strategic goals or the needs of a quickly evolving business environment frequently serves as the catalyst for this process. Agile leaders play a critical role in facilitating cultural shifts inside their firms, leading them through a metamorphosis that improves flexibility, creativity, and efficiency in general.

Leadership Influence and Alignment: Leaders have a pivotal role in initiating cultural transformation by serving as agents of change. Agile leaders understand how important it is to have an impact on corporate culture and match it with strategic objectives. Agile leaders create a compelling vision for the desired culture and connect it with the overall goal through clear actions and continuous communication. Leadership behaviors set the tone for the entire organisation and lay the groundwork for significant cultural transformation by bringing ambitions and behaviours into alignment.

Inclusive Communication and Employee Engagement: All members of the organisation must be involved in inclusive communication for cultural transformation to be effective. Agile leaders place a high value on open and honest communication because it promotes a feeling of understanding and a common goal. They aggressively seek feedback from staff members at all levels, promoting communication and appreciating a range of viewpoints and keeping in mind the different leadership communication styles on fostering inclusion (Allen & Eby, 2020). Agile leaders provide a forum for collaboration, frequent feedback sessions, and town hall meetings to make sure staff members feel heard and involved in the changing culture. This inclusiveness fosters a sense of ownership and increases employee engagement, so cultural transformation is a team effort rather than a top-down directive.

3.6.2 *Leadership Role in Transforming Organisational Culture to Support Innovation*

An organisation's culture is largely shaped by its leadership, which is even more important when it comes to encouraging innovation. Leaders at all levels must work together to change the company culture in order to foster innovation. This is a conversation on the leadership position in this process of transformation:

Setting a Clear Vision and Purpose: A clear vision and purpose that highlights the value of innovation inside the company must be communicated by leaders. This vision acts as a beacon of light, directing all employees' efforts toward a single objective.

Leaders may create a sense of purpose and inspire staff members to contribute their innovative ideas and solutions by telling a story that is appealing and emphasises the importance of innovation in accomplishing the organisation's objective.

Empowering and Trusting Teams: Teams that are given the freedom to own their work tend to produce their best work. Leaders that provide their people freedom and decision-making power can encourage creativity. Teams are more likely to participate in the innovation process when they believe that their contributions are valued and that they have the freedom to try new things and make judgments. Building an atmosphere where staff members are free to experiment with new ideas without needless obstacles is another aspect of building trustworthy teams.

Recognising and Rewarding Innovation: Rewards and recognition are effective strategies for reiterating desired actions. It is important for leaders to set up procedures that recognise and honor creative efforts. This could entail performance incentives, official recognition programs, or even an unofficial culture of gratitude. Leaders encourage employees to actively contribute to the innovation agenda by reinforcing the value of innovation as a fundamental value inside the firm and linking rewards and recognition to inventive initiatives and need to design work environments that foster intrinsic motivation for innovation (Amabile, 2020).

Aligning Organisational Systems and Processes: The systems and procedures of the company must be in line with the leadership's objective of promoting innovation. This covers project management techniques, resource allocation, and performance evaluation measures. Employee prioritisation and effort allocation toward promoting innovation are more likely when these systems encourage and recognise creative endeavors. To make sure that organisational structures are supportive of an innovative culture, leaders must constantly evaluate and modify them.

Encouraging Collaboration and Cross-Functional Teams: Collaboration amongst people with diverse backgrounds frequently fosters innovation. By actively promoting cross-functional cooperation, leaders may dismantle organisational silos. Leaders create an environment that is conducive to the flow of ideas and the discovery of novel solutions by cultivating a convergence of varied perspectives. Creating real and virtual locations that promote communication and idea exchange among team members is another way to promote collaboration.

Measuring and Communicating Progress: Proficient leaders recognise the significance of both assessment and communication in the process of cultural development. They create key performance indicators (KPIs) to measure the innovation efforts of the company and notify all employees on a regular basis of the advancements made. This openness strengthens the organisation's commitment to innovation and fosters responsibility. Leaders build a narrative that maintains momentum and promotes an ongoing emphasis on innovation by sharing success stories and lessons gained.

Adapting to Change and Embracing Disruption: Leaders must be flexible and at ease with change since innovation frequently entails upending established methods of thinking and doing. They ought to exhibit fortitude in the face of ambiguity and accept disruption as a necessary component of the creative process. Agile change managers provide a model for the entire company, encouraging a mindset that sees disturbance as a chance for development and transformation rather than as a danger.

3.7 Conclusion

Leadership is becoming more and more important as the business environment changes and acts as a spark to spur innovation. In-depth analysis of the complex interrelationship between innovation and leadership is provided in this conceptual piece, along with tips, tactics, and case studies to encourage leaders to become agents of revolutionary change. Businesses find the fuel to ignite the road toward a future characterised by innovation and sustainable success in the furnace of visionary leadership.

These accelerators for leadership have one thing in common: they can see alternatives beyond the current state of affairs. Indra Nooyi's emphasis on sustainability and healthier products shows that she is dedicated to innovation that meets the requirements of the environment and society. Jack Ma's innovative work in digital banking and e-commerce has changed the face of international business. These executives have sparked innovation inside their companies with their distinct methods and astute choices. They have developed societies that value adaptation, taking chances, and ongoing development. These catalysts for leadership, whether they be technological innovations, well-timed acquisitions, or cultural transformations, have created the conditions for long-term innovation and competitiveness.

To sum up, the notion of a leadership catalyst functions as a guide for establishments within the intricacies of the contemporary corporate landscape. Inspiring, adapting, and stimulating innovation are not just qualities of a great leader; they are transformative forces that help businesses succeed in the future. The lessons learned from these leadership catalysts offer insightful guidance to aspiring leaders who aim to spur innovation and promote constructive change in their companies as the business landscape changes further. Innovation is a continuous process, and the legacy of leaders who have sparked innovation can inspire upcoming leaders to take on new challenges, see opportunities, and drive their companies into unexplored success areas.

Highlights

- Asian visionary leaders can drive innovation by fostering creativity, investing in infrastructure, and aligning innovation with strategic goals.
- Asian leaders can sharpen their agility by embracing change, promoting collaboration, and encouraging experimentation.

- Asian resilient leaders can navigate setbacks through adaptability, communication, and a growth mindset.
- Asian leadership should continuously evolve and transform organisational culture to sustain innovation through vision, empowerment, and continuous learning.

References

Abraham, C., Nair, K., & Mtegha, A. (n.d.). *Samsung Innovation and Technology*. Research Gate. Retrieved August 17, 2024, from https://www.researchgate.net/publication/343721698_SAMSUNG_INNOVATION_AND_TECHNOLOGY/citations

Afsar, B., Zhang, Z., & Li, R. (2020). Leadership and individual learning agility: A review and research agenda. *Journal of Management Development, 39*(8), 1006–1028.

Allen, T. D., & Eby, L. T. (2020). Leader communication and diversity: Charting a path to inclusion. *Academy of Management Review, 45*(2), 236–262.

Amabile, T. M. (2020). The future of work and the power of intrinsic motivation. *Journal of Applied Behavioral Science, 56*(2), 150–160.

Ashley Lobo. (2020). *Case Study of Toyota Prius—The world's most popular Hybrid Electric Vehicle* |. LinkedIn. https://www.linkedin.com/pulse/case-study-toyota-prius-worlds-most-popular-hybrid-electric-lobo/

Bass, B. M., & Riggio, R. E. (2006). *Transformational Leadership*. Psychology Press.

Biocon Integrated Annual Report (2023–2024). https://www.biocon.com/investor-relations/financial-information/annual-reports/

Bloomberg. (2024). *SoftBank founder Masayoshi Son ready for next big tech bet after hiatus | World News—Business Standard*. https://www.business-standard.com/world-news/softbank-founder-masayoshi-son-ready-for-next-big-tech-bet-after-hiatus-124062000237_1.html

Bunce, P., & Roos, G. (2022). Agile leadership: Navigating resistance and enabling innovation in turbulent environments. *Journal of Organizational Change Management, 35*(6), 1032–1051.

Carmeli, A., Dutton, J. E., & Hardin, A. E. (2021). The relational bases of organizational resilience: A study of interorganizational teams. *Journal of Management, 47*(5), 1084–1115.

Chen, Y., Liu, Y., & Li, S. (2021). Differentiation strategy and firm performance: Moderating effects of market dynamism and environmental turbulence. *International Journal of Business Environment, 14*(3), 130–147.

Chuang, F., Liu, H., & Wu, J. (2021). Leadership in the digital era: Challenges and opportunities. *Journal of Asian Business Research, 24*(3), 45–60.

CTRA Annual-Report. (2023). https://ciputradevelopment.com/investor-relations/

Davis, B., King, D., & Gregory, R. (2021). Developing adaptive expertise in leaders: A multi-method study. *The Leadership Quarterly, 32*(2), Article 101527.

Deci, E. L., & Ryan, R. M. (1987). The support of autonomy and the control of behavior. *Journal of Personality and Social Psychology, 53*(6), 1024–1037.

DeRue, D. S., Ashford, S. J., & Meyers, R. H. (2021). Fostering future-oriented leadership: A roadmap for developing "digital-age" leaders. *Human Resource Management Journal, 31*(2), 309–328.

Dinh, H. T., Nguyen, T. T., & Tran, T. M. (2021). The mediating role of psychological capital in the relationship between transformational leadership and organizational citizenship behavior. *Journal of Global Operations and Marketing, 2*(1), 3–12.

Dhirubhai Ambani | Biography, Accomplishments, and Facts |. (2024). Britannica Money. https://www.britannica.com/money/Dhirubhai-Ambani

Earley, P. C., & Ang, S. (2003). *Cultural Intelligence: Individual Interactions Across Cultures.* Stanford Business Books.

Epstein, M. J., & Buono, A. (2020). Measuring corporate social responsibility: Past, present, and future challenges. *Business Ethics Quarterly, 30*(4), 415–438.

Global. Rakuten (2023). https://global.rakuten.com/corp/CorporateGovernanceReport/2023

Gong, Y., Liu, M., & Wang, Y. (2020). Leader ethical behavior and employee creativity: The mediating role of psychological safety. *International Journal of Manpower, 41*(8), 1308–1327.

Heathfield, S. M. (2021, May 1). What is employee empowerment and how to achieve it? *The Balance Careers.* https://www.thebalancecareers.com/what-is-employee-empowerment-1918144

Heifetz, R. A., & Linsky, M. (2020). The work we must do together: Leading through complexity and paradox. *Journal of Applied Behavioral Science, 56*(2), 161–178.

Hofstede, G. (1980). *Culture's Consequences: International Differences in Work-Related Values.* Sage.

House, R. J., et al. (2004). *Culture, Leadership, and Organizations: The GLOBE Study of 62 Societies.* Sage.

Huawei Technologies Annual Report. (2023). https://www.huawei.com/en/annual-report/2023

Infosys—Consulting | IT Services | Digital Transformation. (n.d.). Retrieved August 17, 2024, from https://www.infosys.com/

Klein, G. (2023). The importance of rapid decision-making in business: A cognitive perspective. *Journal of Decision Sciences, 48*(1), 23–36. https://doi.org/10.1234/jds.2023.001

Kumar, S., & Sharma, A. (2023). Building resilience in innovation: Strategies for sustainable success in dynamic environments. *Journal of Innovation and Knowledge, 8*(1), 103–114.

Li, H., Wang, T., & Zhao, Y. (2018). Leadership styles in China and their impact on innovation. *Chinese Journal of Management, 15*(4), 67–81.

Ma Huateng. (2024). *Forbes.* https://www.forbes.com/profile/ma-huateng/

Nguyen, M., Tran, L., & Vo, T. (2020). Cultural intelligence and leadership effectiveness in Southeast Asia. *International Journal of Cross-Cultural Studies, 8*(2), 91–105.

Nixon, E., & Noble, A. (2022). Overcoming resistance to agile transformation: Strategies for leaders. *Journal of Business Strategy, 43*(5), 14–23.

Pentina, A., & Taylor, L. (2020). Navigating the era of changing customer expectations: A service-dominant logic perspective. *Journal of Services Marketing, 34*(5), 739–754.

Pereira, D. (2024). *BYD Business Model—How BYD Makes Money?* https://businessmodelanalyst.com/byd-business-model/

Petroliam Nasional Berhad (Petronas). (2023). Annual Report. https://www.petronas.com/media/reports

Rigby, D. K., Elk, S., & Berez, S. (2020). The agile c-suite. *Harvard Business Review, 98*(3), 64–73.

SoftBank Group Report. (2024). Annual Report, https://group.softbank/en/ir/financials/annual_reports

Van Dierdonck, M. B., De Dreu, C. K. W., & Nijstad, B. A. (2020). Vision leadership: Developing shared visions through collective imagination. *The Leadership Quarterly, 31*(3), Article 101447.

Wang, G., Liu, Y., & Wang, H. (2022). Visionary leadership and employee creativity: The mediating role of psychological safety and employee empowerment. *International Journal of Manpower, 43*(8), 1716–1731.

Zhang, X., & Wang, L. (2020). Leadership and innovation in the digital age: Lessons from Chinese firms. *Journal of Technology Management in China, 15*(1), 12–29.

Syriac Nellikunnel Devasia is currently the Dean Faculty of Business, PERDANA University and is expert in education leadership and administration and online education management. He is a Certified Trainer from HRDF, Malaysia. He has chaired more than 5 international conferences. He is the recipient of the Most Fabulous Global Knowledge Management Leaders @ World HRD Congress; and has also received the Blackboard Award for Student, *Malaysia.* His research interests include, marketing, sustainable development and innovation, SDGs, CSR, Quality frameworks, consumer behavior and digital transformation strategies. He is also a journal reviewer to number of international journals and is an external examiner for Ph.D. and DBA programmes to a number of reputed universities. He has supervised a number of Ph.D. students.

Soney Haris is an Associate Professor at Department of Commerce and Management-St. Joseph's College of Commerce, Bengaluru City University. She was associated with universities like Latrobe University-Australia, University of Hertfordshire-UK, Laureate International Universities-USA, INTI-International Universities-Malaysia, University of Wollongong, Bangalore University and Jain University-Bangalore and has published various research papers in the area of Marketing Management, Consumer Behaviour, Education, Ethics, Leadership and General Management. She is also a research supervisor for Ph.D. and M.Phil. in the field of commerce and management and a Certified Professional Teacher in Commerce and Management awarded by the Commission of International Accreditation Organisation (IAO)-USA and also a Certified Positive Psychologist and Certified Mindfulness Trainer.

Chapter 4
Research Propositions for Family Business in Asia: A Systematic Literature Review

Clusters of Future Studies: Corporate Foresight; Past and Futures

Jiyoon An

Abstract This chapter provides a review of the scholarship of family business in Asia with future research propositions. The goal of this chapter is to synthesise existing research with the bibliometric analysis and Theory-Context-Characteristics-Methodology (TCCM) framework and draw research propositions. Future researchers may invite a diversity, equity, and inclusion (DEI)-oriented theory (Proposition 1) and a consumer-facing theory (Proposition 2). Research contexts can be broadened to include family-run business entities (e.g., foundation) other than a business (Proposition 3) and the dark side of the family business (Proposition 4). Further investigation can be dedicated to examining the variances among family business clusters grounded in different cultures as business performance drivers (Proposition 5) and the formation and development of differences from process-oriented approaches (Proposition 6) with more sophisticated study designs, including unstructured (Proposition 7) and longitudinal (Proposition 8). This contribution aims to advance scholarship by identifying research gaps and establishing consensus on research priorities.

Keywords Family business · Asia · Bibliometric analysis · Research propositions · TCCM framework

---

J. An (✉)
Broadwell College of Business and Economics, Fayetteville State University, Fayetteville, USA
e-mail: jan@uncfsu.edu

M. S. Balaji et al. (eds.), *Future of Business in Asia*,
https://doi.org/10.1007/978-3-031-95940-0_4

4.1 Introduction

Family businesses are central to understanding the past, present, and future of business landscapes. EY (2023) has celebrated the crucial roles of family businesses: Ford, the 5th largest family business by revenue, forwards consumers mobility choices, and LG, the 18th largest family business by revenue, makes our life good with various consumer electronics. The report also documented the rise of family businesses based in Asia as the region has provided almost one fifth of the largest family businesses in the world.

The scholarly tradition of family business research has been established with the existing family business establishments through industrialisation and colonialisation by examining the Western-centric family business practices and their grounded values (e.g., Judeo-Christian values) (Idemudia, 2011). This Western-centric perspective has slowed the growth of region (e.g., Asia)-specific understanding of family businesses by investigating their pathfinding processes for success and strategic choices grounded in their unique cultural beliefs and systems (Blodgett et al., 2011; Weidenbaum, 1996).

Scant scholarly attention has been dedicated to examining the evolving scholarship of family business in Asia as a meso-level theory to bridge the firm-level micro theory to generalisation-seeking macro theory (Homer & Lim, 2024). This chapter aims to investigate how scholarly narratives advance our understanding of how geography-grounded culture would affect the governance, strategy, and succession of family businesses by using the bibliometric analysis (Donthu et al., 2021; Paul et al., 2021) and Theory-Context-Characteristics-Methodology (TCCM) framework (Bhukya & Paul, 2023) to synthesise the scholarship of family business in Asia and provide future research directions.

4.2 Review Methodology

The bibliometric analysis and Theory-Context-Characteristics-Methodology (TCCM) framework analysis were conducted, followed by the assembly, arrangement, and assessment method for the systematic literature review (Paul et al., 2021; Zupic & Čater, 2015). The bibliometric analysis helps organise prominent academic publications and papers, discovering emerging themes through keyword co-occurrence analysis, and clustering analysis based on backward-looking (such as co-citation) and forward-looking (such as bibliographic coupling) perspectives (Donthu et al., 2021). The TCCM framework provides a comprehensive summary of existing research and draws future directions for theoretical and empirical research from both theory-discovery and theory-testing perspectives (Palmatier et al., 2018; Paul & Benito, 2018). This methodological rigor was designed to answer the following research questions.

RQ1: How can the bibliometric analysis inform the evolving scholarship on family business in Asia?

RQ2: What are the propositions for the future research directions drawn from the TCCM framework?

The assembly, arrangement, and assessment methods were conducted for empirical rigor. For the assembly phase, the study design was determined by the research questions, which selected the keyword and database for data gathering. The keyword strings were selected to gather data for the research questions. The string of (family business OR family-owned OR family-operated OR family firm) AND Asia was used as the search keyword to consider similar terms in the evolving scholarship. The selected search keyword reflected the various terms for family business from the scholarly tradition in a context to bring novel insights to the existing literature for *Theory Adaptation* (Jaakkola, 2020). The Scopus database (e.g., 40,000+ journals) was selected as it covers a more expansive coverage compared to the Web of Science (e.g., 21,000+ journals) database (Clarivate, 2023; Scopus, 2023).

For the arrangement phase, 198 documents (1994–2023) were retrieved from the search, after screening out irrelevant search outcomes (e.g., family planning). The citation analysis was conducted for the publication trend and most influential papers with programming language R with Bibliometrix library. For the assessment phase, data visualisation and interpretation occurred, which was guided by Lim and Kumar (2024). The references were mapped to identify the relationship across the literature from the prospective and retrospective perspectives with the VoS Viewer. The research synthesis and propositions for future research were drawn by the TCCM framework analysis.

4.3 Findings

Figure 4.1 represents the growing scholarly interests over time, with the time as the x-axis and the publication count as the y-axis. The first paper in the domain of family business in Asia was explicitly grounded in anti-orientalism and examined the relationship between Confucianism and Taiwanese family firms from the ethnographical perspective (Greenhalgh, 1994). The quantitative turn occurred with the rise of governance study with ownership and control data from nine countries (e.g., Hon Kong, Indonesia, Japan, South Korea, Malaysia, Philippines, Singapore, Taiwan, and Thailand) (Claessens et al., 2000), garnering thousands of citations. In the 2010s, scholars investigated the adaptive operational strategies of family firms in practice (e.g., resilience and corporate social responsibility) (Amann & Jaussaud, 2014; El Ghoul et al., 2016). R&D and internationalisation have gained more scholarly attention in the 2020s, which may be aligned with the real-world family business practices (Lin & Wang, 2021; Zahra, 2020).

Table 4.1 shows the most influential papers in the scholarship. Claessens and colleagues (2000) was the most cited paper, which may help ignite broad scholarly

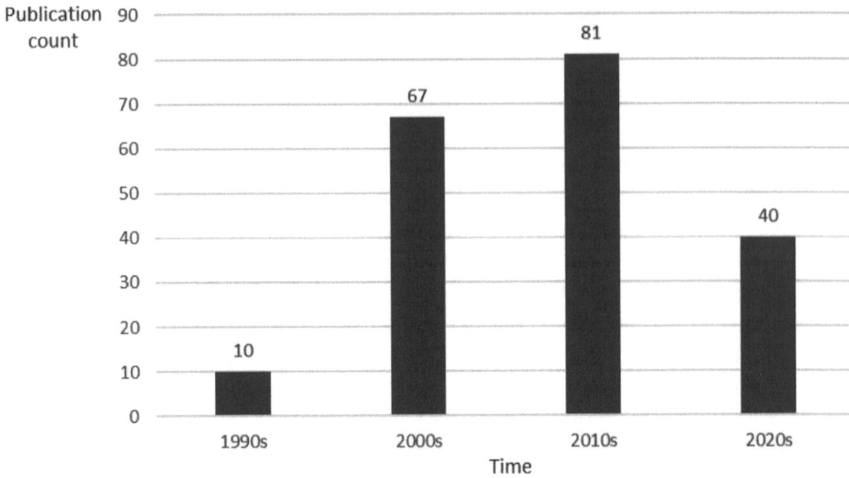

Fig. 4.1 Publication trend of the scholarship in family business in Asia

interests. Regionally bounded entrepreneurship (e.g., East Asia, China) is popular contexts among them to investigate its unique management and operation practices (e.g., ownership, risk management, and stakeholder management) compared to the Western equivalents (Au & Kwan, 2009; Carney, 1998; Mishra, 2011; Walter, 2011; Wong & Ng, 2002).

Figures 4.2 and 4.3 indicate the retrospective and prospective perspectives to organise research streams. Whereas the established research streams from the backward-looking view are theoretically grounded (e.g., agency theory, transaction cost theory, resource-based view, and social capital theory), the emerging research streams adopt the practical perspectives to examine the family firm's financial and non-financial involvements and performances and their strategic choices.

Figure 4.2 identified four clusters: Ownership and control (red cluster), agency theory perspective (green cluster), family embeddedness and competitive advantage (yellow cluster), and social capital (blue cluster).

Figure 4.3 recognised six clusters: leadership practice (yellow cluster), firm performance assessment (purple cluster), ownership and control strategy (green cluster), family involvement (grey cluster), comparative perspectives (blue cluster) and ethics (red cluster).

Figure 4.4 shows the application of TCCM framework to synthesize existing literature, which was guided by Lim and Kumar (2024)'s scanning, sensing, and substantiating method. As the goal of this research is to enhance a region-based scholarly understanding to complement Western-centric existing literature, the focal topic, family business in Asia, was examined via multi-level perspectives (e.g., micro, meso, and macro) for Theory Adaptation (Jaakkola, 2020).

From the micro theoretical perspective, how family funding reduces a startup's transaction cost to raise financial resources was examined with the transaction cost

Table 4.1 Most influential papers in the scholarship on family business in Asia

Title	Author	Year	Total citations
The separation of ownership and control in East Asian corporations	Claessens et al.	2000	3180
A study of the relationship between corporate governance structures and the extent of voluntary disclosure	Ho and Wong	2001	637
Governing finance: East Asia's adoption of international standards	Walter	2008	211
Restructuring industrial districts, scaling up regional development: a study of the Wenzhou model, china	Wei et al.	2007	169
Are family ownership and control in large firms good, bad, or irrelevant?	Jiang and Peng	2011	152
Family control and corporate social responsibility	El ghoul et al.	2016	135
Guidelines for family business boards of directors	Lane et al.	2006	132
De-orientalising the Chinese family firm	Greenhalgh	1994	127
Familial capitalism in global institutional contexts: implications for corporate governance and entrepreneurship in East Asia	Steier	2009	106
Start-up capital and Chinese entrepreneurs: the role of family	Au and Kwan	2009	105
The stranded individualiser under compressed modernity: South Korean women in individualisation without individualism	Chang and Song	2010	104
A management capacity constraint? Obstacles to the development of the overseas Chinese family business	Carney	1998	104
The emergence of small transnational enterprise in Vancouver: the case of Chinese entrepreneur immigrants	Wong and Ng	2002	103
Family and non-family business resilience in an economic downturn	Amann and Jaussaud	2012	99
Multiple large shareholders and corporate risk taking: evidence from East Asia	Mishra	2011	90

theory perspective (Au & Kwan, 2009). Scholars have investigated family firms' agency problems, which worsened corporate social responsibility in countries with weak institutions (El Ghoul et al., 2016). From the meso perspective, scholars have examined how family ties can foster the internationalisation due to the pooled resources based on the resource-based view (Chung & Dahms, 2016). Social capital theory was examined in the context of Central Asia on how family firms bring advantages of inter-generational business operations in recruitment, training, and employee career development (Roberts et al., 2009). Institutional theory was used to investigate the meso-level phenomena to compare the Western and Asian private family firm's finance operation strategies (Doucet et al., 2022).

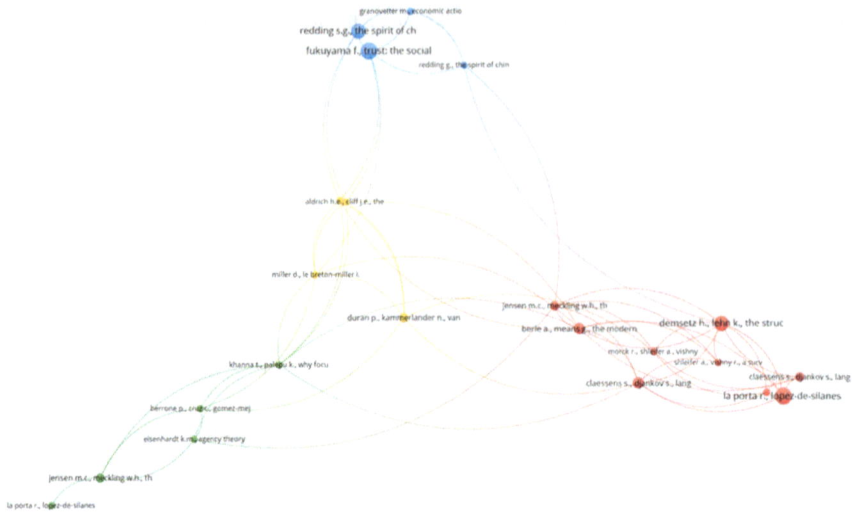

Fig. 4.2 Co-citation analysis of the scholarship in family business in Asia

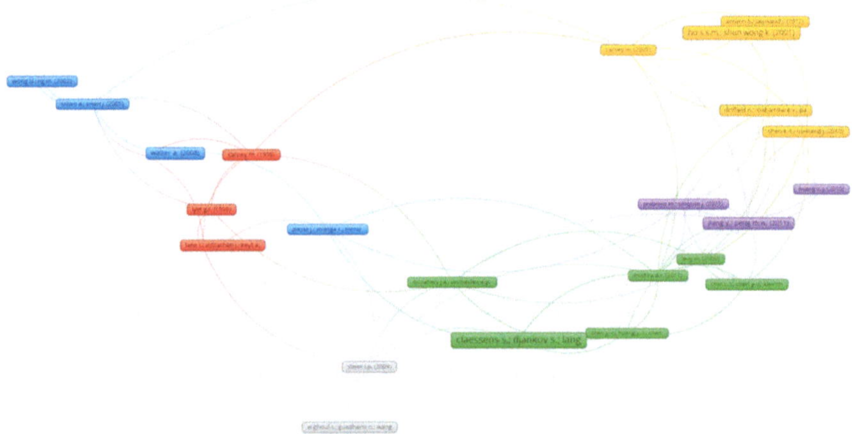

Fig. 4.3 Bibliographic coupling analysis of the scholarship in family business in Asia

Succession (Chung & Au, 2021), regional development (Kienzle & Shadur, 1997), innovation (Ding et al., 2021), and market development (Carney, 1998) were frequently researched contexts. Scholars have been interested in researching the characteristics of leadership (Qureshi et al., 2021), ownership (McCahery & Vermeulen, 2010), firm performance (Her & Mahajan, 2005), and social reasonability (Fernando & Almeida, 2012) of family firms in Asia. Existing literature has utilised inductive/qualitative (e.g., interview) and deductive/quantitative (e.g., survey) in the cross-sectional research design.

Fig. 4.4 Theory-context-characteristics-methodology (TCCM) analysis for the scholarship of family business in Asia

4.4 Future Directions

The TCCM (Theory-Context-Characteristics-Methodology) framework helped implement the MacInnis (2011) method to envision, explicate, relate, and debate the scholarship of existing research for future research directions. Research propositions for future research are summarised in Table 4.2.

The theory dimension of the scholarship can be advanced to invite a diversity, equity, and inclusion (DEI)-oriented theory (Proposition 1) and a consumer-facing theory (Proposition 2). Future research may investigate how family business either fosters or hinders the impact of intersectionality on social mobility (Qureshi et al., 2023; Romero & Valdez, 2016) from marginalised groups in the context of entrepreneurship in the context of Asia. As Asia covers a wide range of belief systems across countries to situate a marginalised social group, it would be worthy of empirical research on the role of family business in social justice. Scholarly attention has

Table 4.2 Research propositions for the scholarship in family business in Asia

Theory	Context	Characteristics	Method
Proposition 1. Diversity, equity, and inclusion (DEI)-oriented theories (e.g., intersectionality) should be included **Proposition 2.** Consumer-facing theory must be included	**Proposition 3.** Family-run business entities (e.g., foundation) needs to be researched **Proposition 4.** Dark side of family business requires further examination	**Proposition 5.** Beyond the firm level analysis should be considered **Proposition 6.** Process-oriented approaches must be adopted to investigate the evolution	**Proposition 7.** Unstructured data should be considered for future research **Proposition 8.** Longitudinal research design needs to be implemented

been dedicated to market responses to family firms' signalling strategies on family values (Bargoni et al., 2023; Galvagno et al., 2023). Further research may investigate what theoretical understanding would advance how consumers interpret family values and family embeddedness in their interactions across various products (e.g., wine) or services (e.g., historic hotels).

The context dimension may pay more attention to more diverse types of family-run business entities (e.g., foundation) other than a business (Proposition 3) and the dark side of the family business (Proposition 4). Scholars have noted the importance of family charitable giving and how it helps maintain family values and build relationships with the community (Feliu & Botero, 2016). Since family charitable giving is an emerging phenomenon compared to the mature one in the West, future research may investigate it to build the typology and identify antecedents, moderators, and outcomes in the context of Asia. The contribution of the family business to entrepreneurial ventures is not without limitations. Scholars have reported the dysfunctional conservatism to prevent revitalisation in the crisis (Montiel Mendez & Soto Maciel, 2021). Future research may use a case study to contextualise how the dark side affects family business operations and strategies, generating insights for healthy family businesses.

Future researchers may investigate the characteristics of family firms in Asia by focusing on the variances among family business clusters grounded in different cultures as business performance drivers (Proposition 5) and the formation and development of differences from the process-oriented approaches (Proposition 6). Existing literature viewed governance, resource, and goal dimensions of family businesses as either an antecedent or a consequence of family business performance, with the firm as the unit of analysis (Fang et al., 2022). Future research may adopt the ecosystem approach by investigating the family business entrepreneurial ecosystem (De Massis et al., 2021) as the unit of analysis, which helps understand the formation, development, and demise of evolving family businesses to contextualize resilience and family values. This perspective is relevant to the current status of family businesses adapting to the service economy to encourage family businesses to function as an evolving service ecosystem.

Future research may implement a more sophisticated study design with unstructured (Proposition 7) and longitudinal (Proposition 8). Digital transformation and big data make artifacts of family businesses readily available as unstructured data (e.g., images, video, and text). Natural language processing and other machine learning techniques can be applied to understand historical archives of unstructured data to analyse practices of family businesses in a longitudinal context.

4.5 Conclusion

This book chapter aims to synthesise existing research with the bibliometric analysis and TCCM framework and draw research propositions for further research. The results help enhance our understanding of the evolving scholarship on family business in Asia and develop propositions for future research directions.

Future researchers may invite a diversity, equity, and inclusion (DEI)-oriented theory (Proposition 1) and a consumer-facing theory (Proposition 2). Research contexts can be broadened to include family-run business entities (e.g., foundations) other than a business (Proposition 3) and the dark side of the family business (Proposition 4). Further investigation can be dedicated to examining the variances among family business clusters grounded in different cultures as business performance drivers (Proposition 5) and the formation and development of differences from the process-oriented approaches (Proposition 6) with more sophisticated study designs with unstructured (Proposition 7) and longitudinal (Proposition 8).

The suggested research propositions may allow for implementing Theory Adaptation (Jaakkola, 2020) to bring novel insights with a regionally grounded meta-context to the Western-centric existing scholarship (Homer & Lim, 2024). The advancement of scholarship on family business in Asia would underscore the economic transformation recipes with indigenous cultures, which helps nurture economic growth and greater well-being beyond Asia. This approach would create scientific, business, and societal impacts by fostering scholarly discourses, addressing business challenges, and supporting sustainability initiatives for practitioners and policymakers (Jaakkola & Vargo, 2021).

Highlights

- Assess the scholarship on family business for future research propositions
- Provide the intellectual foundations for evidence-based research
- Present non-Western centric perspectives for more inclusive business leadership.

References

Amann, B., & Jaussaud, J. (2014). Family and non-family business resilience in an economic downturn. In *Japan's' Lost Decade'* (pp. 69–89). Routledge.

Au, K., & Kwan, H. K. (2009). Start–up capital and Chinese entrepreneurs: The role of family. *Entrepreneurship Theory and Practice, 33*(4), 889–908.

Bargoni, A., Alon, I., & Ferraris, A. (2023). A systematic review of family business and consumer behaviour. *Journal of Business Research, 158*, Article 113698.

Bhukya, R., & Paul, J. (2023). Social influence research in consumer behavior: What we learned and what we need to learn?–A hybrid systematic literature review. *Journal of Business Research, 162*, Article 113870.

Blodgett, M. S., Dumas, C., & Zanzi, A. (2011). Emerging trends in global ethics: A comparative study of US and international family business values. *Journal of Business Ethics, 99*, 29–38.

Carney, M. (1998). A management capacity constraint? Obstacles to the development of the overseas Chinese family business. *Asia Pacific Journal of Management, 15*, 137–162.

Chung, H.-M., & Au, K. (2021). Why succession is a critical issue for SMEs: Perspectives and cases in the Asian context. *Succession and Innovation in Asia's Small-and-Medium-Sized Enterprises*, 1–25.

Chung, H.-M., & Dahms, S. (2016). Particularistic ties and internationalisation: Evidence from Taiwanese family business groups. *European Journal of International Management, 10*(5), 558–580.

Claessens, S., Djankov, S., & Lang, L. H. P. (2000). The separation of ownership and control in East Asian corporations. *Journal of Financial Economics, 58*(1–2), 81–112.

Clarivate. (2023). "Web of Science Platform", available at: https://clarivate.libguides.com/webofscienceplatform/woscc. Assessed December 5, 2023.

De Massis, A., Kotlar, J., & Manelli, L. (2021). Family firms, family boundary organizations, and the family-related organizational ecosystem. *Family Business Review, 34*(4), 350–364.

Ding, H.-b., Chung, H.-M., Yu, A., & Phan, P. H., (Eds.). (2021). *Innovation, Growth, and Succession in Asian Family Enterprises*. Edward Elgar Publishing.

Donthu, N., Kumar, S., Mukherjee, D., Pandey, N., & Marc Lim, W. (2021). How to conduct a bibliometric analysis: An overview and guidelines. *Journal of Business Research, 133*, 285–296.

Doucet, P., Requejo, I., & Suárez-González, I. (2022). Diversification decisions of family SMEs under uncertainty: Goals as a rule of thumb. *Entrepreneurship Research Journal, 0*.

El Ghoul, S., Guedhami, O., Wang, H., & Kwok, C. C. Y. (2016). Family control and corporate social responsibility. *Journal of Banking & Finance, 73*, 131–146.

EY. (2023). How the World's Largest Family Businesses are Outstripping Global Economic Growth, available: https://familybusinessindex.com/. Accessed October 23, 2024.

Fang, H., Chevy, K. S., Kim, T., Marler, L., & Chrisman. J. J. (2022). Family business research in Asia: Review and future directions. *Asia Pacific Journal of Management*, 1–42.

Feliu, N., & Botero, I. C. (2016). Philanthropy in family enterprises: A review of literature. *Family Business Review, 29*(1), 121–141.

Fernando, M., & Almeida, S. (2012). The organizational virtuousness of strategic corporate social responsibility: A case study of the Sri Lankan family-owned enterprise MAS Holdings. *European Management Journal, 30*(6), 564–576.

Galvagno, M., Pisano, V., & Strano, S. M. (2023). Family business branding from a signaling theory perspective: An integrative framework. *Journal of Product & Brand Management, 32*(5), 681–696.

Greenhalgh, S. (1994). De-orientalizing the Chinese family firm. *American Ethnologist, 21*(4), 746–775.

Her, M. M., & Mahajan, A. (2005). Family control, two-tier boards and firm performance: Lessons from the Taiwanese experience. *Journal of Asia-Pacific Business, 6*(2), 69–89.

Homer, S. T., & Lim, W. M. (2024). Theory development in a globalized world: Bridging "doing as the romans do" with "understanding why the romans do it. *Global Business and Organizational Excellence, 43*(3), 127–138.

Idemudia, U. (2011). Corporate social responsibility and developing countries: Moving the critical CSR research agenda in Africa forward. *Progress in Development Studies, 11*(1), 1–18.

Jaakkola, E. (2020). Designing conceptual articles: Four approaches. *AMS Review, 10*(1), 18–26.

Jaakkola, E., & Vargo, S. L. (2021). Assessing and enhancing the impact potential of marketing articles. *AMS Review, 11*(3), 407–415.

Kienzle, R., & Shadur, M. (1997). Developments in business networks in East Asia. *Management Decision, 35*(1), 23–32.

Lim, W. M., & Kumar, S. (2024). Guidelines for interpreting the results of bibliometric analysis: A sensemaking approach. *Global Business and Organizational Excellence, 43*(2), 17–26.

Lin, W. T., & Wang, L. C. (2021). Family firms, R&D, and internationalization: The stewardship and socio-emotional wealth perspectives. *Asia Pacific Journal of Management, 38*(1), 91–119.

MacInnis, D. J. (2011). A framework for conceptual contributions in marketing. *Journal of Marketing, 75*(4), 136–154.

McCahery, J. A., & Vermeulen, E. P. M. (2010). *Corporate governance of non-listed companies.* Oxford University Press.

Mishra, D. R. (2011). Multiple large shareholders and corporate risk taking: Evidence from East Asia. *Corporate Governance: An International Review, 19*(6), 507–528.

Montiel Mendez, O. J., & Soto Maciel, A. (2021). Dark side of the family business: an exploratory perspective. *Journal of Family Business Management, 11*(4), 386–401.

Palmatier, R. W., Houston, M. B., & Hulland, J. (2018). Review articles: Purpose, process, and structure. *Journal of the Academy of Marketing Science, 46*, 1–5.

Paul, J., & Benito, G. R. G. (2018). A review of research on outward foreign direct investment from emerging countries, including China: What do we know, how do we know and where should we be heading?. *Asia Pacific Business Review, 24*(1), 90–115.

Paul, J., Lim, W. M., O'Cass, A., Wei Hao, A., & Bresciani, S. (2021). Scientific procedures and rationales for systematic literature reviews (SPAR-4-SLR). *International Journal of Consumer Studies, 45*(4), O1–O16.

Qureshi, I., Bhatt, B., Sutter, C., & Shukla, D. M. (2023). Social entrepreneurship and intersectionality: Mitigating extreme exclusion. *Journal of Business Venturing, 38*(2), 106283.

Qureshi, J. A., Shaikh, A. M., & Seaman, C. (2021). Leadership mindset and the fall of once giant family-run textile exporting businesses. *Global Business and Organizational Excellence, 40*(6), 41–55.

Roberts, K., Kamruzzaman, P., & Tholen, J. (2009). Young people's education to work transitions and inter-generational social mobility in post-soviet central Asia. *Young, 17*(1), 59–80.

Romero, M., & Valdez, Z. (2016). Introduction to the special issue: Intersectionality and entrepreneurship. *Ethnic and Racial Studies, 39*(9), 1553–1565.

Scopus. (2023). "Scopus preview," available at: https://www.scopus.com/sources. Assessed December 5, 2023.

Walter, A. (2011). *Governing finance: East Asia's adoption of international standards.* Cornell University Press.

Weidenbaum, M. (1996). The Chinese family business enterprise. *California Management Review, 38*(4), 141–156.

Wong, L. L., & Ng, M. (2002). The emergence of small transnational enterprise in Vancouver: The case of Chinese entrepreneur immigrants. *International Journal of Urban and Regional Research, 26*(3), 508–530.

Zahra, S. A. (2020). Technological capabilities and international expansion: The moderating role of family and non-family firms' social capital. *Asia Pacific Journal of Management, 37*(2), 391–415.

Zupic, I., & Čater, T. (2015). Bibliometric methods in management and organization. *Organizational Research Methods, 18*(3), 429–472.

Jiyoon An (Ph.D., University of Rhode Island) is an Assistant Professor of Marketing at Fayetteville State University, NC, USA. Educated in South Korea, France, and the USA, her research has a global perspective and has appeared in peer-reviewed journals, including the *Social Responsibility Journal, Consumer Behavior in Tourism and Hospitality, Maritime Business Review*, and *Journal of Business Analytics*. Her research lies at the intersection of entrepreneurship, marketing, innovation, and data analytics.

Chapter 5
Women in Business: A Historical Perspective and Future Outlook in Malaysia

Clusters of Future Studies: Past and Futures

Hemavathi Gunasegaran

Abstract This chapter explores the historical evolution, present challenges, and future prospects of women's participation in Malaysia's business sector. As a review paper, it draws insights from a synthesis of policy documents, academic literature, and industry reports. By adopting a qualitative desk research approach, the chapter examines significant milestones, such as the New Economic Policy (1971) and the National Policy on Women (1989), alongside trends in labour force participation, entrepreneurship, and leadership roles. The methodology involves a thematic analysis of secondary data sources to uncover structural barriers, cultural norms, and policy-driven impacts that shape women's roles in business. Through this lens, the chapter identifies key enablers and challenges, emphasising the role of digital transformation, targeted policy reforms, and capacity-building initiatives in promoting gender inclusivity. The findings emphasise the necessity of targeted policy reforms, digital literacy programs, and educational efforts to bridge gender gaps in business participation. Future directions focus on fostering inclusivity through supportive networks, enhancing rural women's entrepreneurship, and leveraging technology for equitable growth. The chapter concludes with actionable recommendations for policymakers, educators, and industry leaders to ensure sustainable progress toward Malaysia's National Entrepreneurship Policy (DKN) in 2030.

Keywords Women in business · Gender inclusivity · Entrepreneurship · Policy reforms · Digital transformation

H. Gunasegaran (✉)
Centre for Continuing Education, Universiti Malaya, Cyberjaya 63000, Selangor, Malaysia
e-mail: hemavathi@um.edu.my

M. S. Balaji et al. (eds.), *Future of Business in Asia*,
https://doi.org/10.1007/978-3-031-95940-0_5

5.1 Introduction

Since gaining independence in 1957, Malaysia has undergone a remarkable economic transformation, shifting from an agrarian-based economy to one driven by manufacturing and services (World Bank, 2020). This evolution, however, has not been without challenges, particularly during the COVID-19 pandemic, which caused widespread disruptions. Nevertheless, one notable development has been the increasing participation of women in business. Specifically, statistics from DOSM (2022) indicate that 187,265 establishments are owned by women, contributing 20.3% to the national economic sector. To further empower these entrepreneurs, the Malaysian government allocated RM241.25 million (approximately USD 55.25 million) in Budget 2020. Thus, the role of women in Malaysia's economy cannot be overstated, as their contributions reflect both economic progress and societal advancements toward gender inclusivity.

Moreover, women-owned small and medium enterprises (SMEs) play an integral role in Malaysia's economic framework. Not only do SMEs contribute around 41% of the country's GDP, but they also provide approximately 70% of employment opportunities (Ghazali et al., 2020; Hassan et al., 2019). Despite this significant impact, women entrepreneurs continue to face numerous challenges, such as financial constraints, sociocultural barriers, and limited representation in leadership. For example, as of 2018, women accounted for only 15.4% of top management positions in Malaysia's top 100 publicly listed companies (Ahmad et al., 2020). These disparities highlight the urgent need for targeted strategies to address gender gaps in business.

On a global scale, the number of women entrepreneurs has steadily risen, particularly in developed countries. Studies underscore their critical contributions to economic growth, innovation, and social progress (Noor et al., 2021; Rashid & Ratten, 2020). However, while global trends provide a valuable backdrop, it is essential to focus on Malaysia's unique context, where historical and cultural factors have profoundly influenced women's participation in business. Therefore, situating this discussion within a Malaysian framework allows for a more nuanced understanding of the challenges and opportunities specific to the country.

Consequently, this study seeks to explore the historical trajectory of women in Malaysia's business sector, from pioneering figures who navigated male-dominated industries to contemporary leaders shaping the future. Specifically, the study aims to:

1. Examine the historical milestones and challenges that have shaped women's participation in Malaysian business.
2. Develop strategies informed by historical insights to foster a more inclusive and equitable business environment.

To achieve these objectives, the paper is structured as follows: it begins with a comprehensive overview of the historical background, followed by an in-depth analysis of findings derived from both secondary and primary sources. Subsequently, the study synthesises these insights and concludes with actionable recommendations. Ultimately, this paper aims to celebrate the legacy of Malaysian women in business while charting a path toward a more inclusive and equitable future.

5.2 Review Methodology

This study adopts a systematic literature review (SLR) methodology to investigate the historical evolution and current status of women's participation in the Malaysian business sector. The SLR approach was chosen to ensure a comprehensive, unbiased, and rigorous review of the existing literature, enabling an in-depth understanding of trends, challenges, and opportunities in this domain. Following the SPAR-4-SLR guidelines outlined by Paul et al. (2021), this methodology integrates diverse sources of information, spanning peer-reviewed academic journals, government reports, and contemporary media articles. Such a multifaceted approach ensures both academic rigor and practical relevance, while providing a broad and contextualised understanding of the subject matter.

The data collection process involved systematically identifying, selecting, and analysing relevant literature. To capture longitudinal trends, the study focused on publications from the 1970s to 2023, emphasising the historical trajectory of women in Malaysian business. The search employed keywords such as women in business, Malaysia, gender equity, entrepreneurship, and policy reforms, using academic databases like Scopus, Web of Science, and Google Scholar. Sources included academic studies focusing on gender studies and business history, government policy documents detailing employment trends and gender equity initiatives, and contemporary reports from non-governmental organisations and business associations to reflect current challenges and opportunities. This combination of sources ensures a robust foundation for exploring the historical and contemporary dynamics of women's business participation in Malaysia.

The study employed the PRISMA framework (Preferred Reporting Items for Systematic Reviews and Meta-Analyses) to streamline the review process. Following this framework, the initial search identified over 500 articles, which were subsequently screened for relevance and duplication, reducing the pool to 150 articles. A full-text review further refined this selection, resulting in a final set of 64 articles for qualitative synthesis. The inclusion criteria prioritised studies that focused on the role of women in Malaysian business, historical milestones, policy interventions, and empirical evidence of progress or challenges. This rigorous screening process ensured the reliability and comprehensiveness of the review.

To enable a critical evaluation of patterns and changes over time, a thematic analysis approach was employed. This method ensured a critical evaluation of the evolution of women's roles, rather than a mere compilation of data, as recommended by Mohamed Shaffril et al. (2021). Key themes, such as policy reforms, societal perceptions, and economic contributions, were identified and analysed across three distinct historical periods: pre-independence (before 1957), post-independence

(1957–1980s), and the modern era (1990s to present). This longitudinal analysis facilitated meaningful comparisons, allowing the study to highlight specific milestones, shifts in societal attitudes, and the impact of policy interventions on women's participation in business. Such an analytical framework aligns with the study's objectives of examining historical milestones and leveraging insights to inform future strategies.

To ensure objectivity and minimise bias, the study incorporated diverse perspectives from various disciplines and sources. By including literature from both academic and non-academic domains, as well as historical and contemporary accounts, the review provides a balanced and holistic understanding of the topic. Furthermore, all interpretations and conclusions are grounded in evidence and supported by citations, avoiding the influence of personal perspectives or unsubstantiated claims.

This approach strengthens the study's credibility while offering a clearer understanding of how women's roles in Malaysia's business sector have evolved over time. By comparing different periods and looking at long-term trends, the study highlights the connections between key historical events, changes in society, and the policies that have shaped women's participation in business. The findings provide useful insights for creating a more inclusive and gender-balanced business environment. Ultimately, the goal is to add to the ongoing conversation about empowering women, while providing practical suggestions for those working to improve women's representation in Malaysia's economy.

5.3 Findings

This study traces the evolution of women's participation in Malaysia's business sector, analysing key milestones, challenges, and emerging opportunities. The journey of promoting women's economic involvement in Malaysia began in 1989 with the formulation of the National Women's Policy (NWP). This policy aimed to ensure equal partnership between men and women, granting women access to resources and opportunities for economic participation (Hussin et al., 2020). As part of the broader commitment to gender equality, the 6th Malaysia Plan (1991–1995) introduced a special fund designed to support women's development, highlighting the early recognition of their contributions to national economic growth (Zaimah et al., 2013).

The 1990s marked a further push to integrate women into the emerging digital economy. The introduction of the Multimedia Super Corridor (MSC) initiative in 1996 provided women with increased access to information and communication technologies (ICT), thus enhancing their skills and participation in the ICT sector (Masrom & Ismail, 2010). These progressive steps continued into the twenty-first century, particularly with the 2011 implementation of a 30% gender quota for decision-making positions, which aimed to improve women's representation in leadership roles across corporate sectors (Elias, 2015). Despite such advancements, however, substantial barriers still exist, particularly in achieving broader gender equity in leadership positions. In 2016, only 17.9% of decision-making roles were held by women in the top 50 public-listed companies in Malaysia, illustrating the continuing struggle for gender parity in leadership (Halimah et al., 2019).

The government's commitment to empowering women was reaffirmed in 2018, when the year was declared "Women Empowerment Year." This declaration was part of a broader national agenda to promote gender inclusivity in various sectors, including corporate leadership and entrepreneurship (Halimah et al., 2019). However, despite these positive strides, traditional gender norms continue to place significant limitations on women's ability to ascend to leadership positions and fully capitalise on entrepreneurial opportunities. Women often face societal expectations that prioritise domestic responsibilities over career advancement, which continues to hinder their professional progress (Bullough et al., 2022).

Another major challenge for women entrepreneurs in Malaysia is their limited access to financial resources, mentorship, and professional networks, which remain significant barriers to their growth and success. According to the Malaysia Gender Gap Report 2024, women in business are still unable to access adequate capital, with only 18% of business financing allocated to female-led enterprises (Chern, 2024). This lack of financial support is compounded by a lack of family support and the cultural expectation that women bear the brunt of household responsibilities, limiting their ability to scale their businesses effectively (Bullough et al., 2022). Furthermore, the intersectionality of gender, ethnicity, and religion creates additional layers of complexity, with marginalised women facing compounded barriers to accessing resources and support (Adapa & Sheridan, 2021).

Despite these barriers, women entrepreneurs have shown remarkable resilience and ingenuity. Across various industries such as retail, manufacturing, and technology, women-led businesses are increasingly contributing to Malaysia's GDP. In 2023, women-led businesses accounted for 18% of the GDP, a notable increase from 12% in 2010 (Maemunah et al., 2023). This increase is attributed to women's growing adoption of digital tools, sustainable business practices, and an increased emphasis on corporate social responsibility and ethical business models (Schwak, 2021). Women entrepreneurs are not only advancing business growth but also fostering innovation and sustainability across sectors.

Government initiatives, such as the Women Entrepreneur Financing Programme, have made strides in increasing access to financial resources. However, limited outreach and low engagement rates have hindered their broader impact (Chipfunde et al., 2021). Programs need to be more inclusive, particularly for women in rural areas and those from marginalised communities, to ensure that all women entrepreneurs have the resources they need to thrive. Additionally, while Malaysia has implemented several policies aimed at increasing gender equality, women continue to face systemic barriers, which limit their full participation in the economy.

The Malaysian government has made significant strides in advancing women's roles in entrepreneurship and leadership. Key milestones in the policy landscape include:

- 1970s: The inclusion of gender equity in national development policies, which emphasised women's role in economic growth (Sakai, 2022).

- 1993: The establishment of the National Association of Women Entrepreneurs of Malaysia (NAWEM), which aimed to support women entrepreneurs and foster collaborative networks (Hussin et al., 2020).
- 2001: The introduction of the National Policy on Women, which created a framework for equitable economic participation (Zaimah et al., 2013).
- 2011: The introduction of a 30% gender quota in corporate decision-making roles, a critical step toward increasing female representation in leadership (Elias, 2015).
- 2018: The declaration of Women Empowerment Year, which underscored the importance of gender inclusivity in sectors such as corporate leadership and entrepreneurship (Halimah et al., 2019).

These milestones demonstrate the country's ongoing commitment to improving gender equity in business. However, despite significant progress, challenges persist. Limited access to financial resources, insufficient mentorship, and gender biases remain as persistent obstacles. The literature highlights that, although progress has been made, cultural expectations continue to present formidable barriers for women in business. Many women entrepreneurs report difficulties in securing capital and expanding their businesses due to both institutional and societal factors (Bullough et al., 2022; Murugiah, 2022).

On a broader scale, the representation of women in leadership roles is still disproportionately low. As of 2022, women hold just 21% of decision-making positions and 10% of board roles in Malaysia's top 50 public-listed companies (Maemunah et al., 2023). In comparison, other countries in the region, such as Singapore and Thailand, have implemented more aggressive policies to close the gender gap in business leadership and entrepreneurship, which Malaysia could learn from to accelerate its progress (Chipfunde et al., 2021).

Looking forward, the digital economy presents an opportunity for women entrepreneurs to expand their businesses, particularly in light of the COVID-19 pandemic, which has further highlighted the importance of technology and digital transformation. Initiatives such as the Women Entrepreneur Financing Programme and various partnerships with non-governmental organisations (NGOs) aim to create an inclusive entrepreneurial ecosystem. However, for sustained growth, Malaysia must address systemic challenges such as unequal access to financial resources, low engagement in government programs, and entrenched cultural norms. These systemic challenges must be tackled through a comprehensive and inclusive approach that integrates policy reforms, enhanced financial support, and a societal shift in gender norms.

In conclusion, while the historical milestones demonstrate Malaysia's commitment to gender equity, substantial barriers remain. Women entrepreneurs continue to face challenges, such as limited access to resources, patriarchal norms, and systemic inequalities. Despite these challenges, women entrepreneurs in Malaysia have shown remarkable resilience and are increasingly contributing to the country's economic growth. Henceforth, addressing these structural and systemic issues through targeted policies and programs, increasing access to capital, and reshaping societal perceptions of gender roles will be crucial in fostering a more inclusive business environment and achieving gender parity in entrepreneurship and leadership.

5.4 Future Direction

Building on the historical insights and analysis of women's contributions to Malaysia's business sector, it is essential to chart actionable strategies for fostering greater gender parity and inclusivity. While significant progress has been made through key policies, such as the National Policy on Women and gender quotas for leadership roles (Abdullah, 2014; Halimah et al., 2019), the practical implementation of these policies is still hindered by structural and cultural barriers. Patriarchal mindsets, workplace resistance, and limited support systems continue to obstruct the full realisation of these initiatives, as seen in the continued underrepresentation of women in leadership positions (Halimah et al., 2019). As Malaysia strives to close the gender gap, future efforts must focus on addressing these challenges while leveraging emerging opportunities in technology and policy reform.

One promising avenue for change lies in embracing digital transformation. Technological advancements, particularly in e-commerce, fintech, and remote work, have already reduced traditional barriers to business participation (Ughetto et al., 2020). These innovations have created new pathways for women entrepreneurs, allowing them to access global markets and grow their businesses. To fully capitalise on these opportunities, future initiatives should prioritise digital literacy programs, tailored training, and skill development that equip women entrepreneurs with the necessary tools to thrive in the digital economy (Abduvakhidov et al., 2021). Special attention must be given to addressing disparities in access to technology, particularly in rural areas, to ensure that all women can participate in the digital economy (Nambisan, 2017).

In addition to digital inclusion, there remains a need for targeted policy reforms that tackle structural barriers hindering women's full participation in business. While Malaysia's gender quotas and policies have made strides, further efforts are required to address pay disparities, leadership gaps, and entrepreneurial support systems (Ariffin et al., 2020; Baqutayan, 2020). It is crucial for policymakers to implement flexible workforce initiatives, enforce anti-discrimination laws, and introduce leadership development programs that provide women with the tools to ascend to top decision-making roles (Halimah et al., 2019). Moreover, fostering entrepreneurship among rural women by providing targeted financial support, mentorship, and infrastructure will be key to driving inclusive growth across Malaysia (Dutkiewicz & Ellis, 2018).

Education remains a cornerstone for empowering women in business. Expanding women's access to business education, particularly in fields such as STEM and sustainability, will strengthen their ability to participate in emerging industries, including those aligned with environmental, social, and governance (ESG) priorities (Mahajan & Bandyopadhyay, 2021). Leadership development programs and capacity-building initiatives specifically designed for women can help close the gender gap in leadership roles, while also addressing the unique challenges faced by women in traditionally male-dominated sectors (Ab Halim, 2023; Kuppuswami & Ferreira, 2022). Societal shifts, challenging traditional perceptions of women's roles in business and leadership, are equally important to creating an environment where women can thrive.

Access to networks and support systems is another critical factor in empowering women entrepreneurs. Women's associations, professional networks, and mentorship programs offer invaluable resources, opportunities for collaboration, and a platform for advocating for gender-sensitive policies (Kyrgidou et al., 2021; Manello et al., 2020). Research has shown that women's involvement in business networks is often limited, particularly in non-traditional sectors, which constrains their ability to grow and scale businesses (Bogren et al., 2013; Sharafizad, 2011). Establishing strong networks will not only enhance access to resources but also build women's confidence to enter and succeed in the business sector.

Continuous research is necessary to inform and refine gender inclusivity strategies in business. Longitudinal studies evaluating the long-term impact of policies and initiatives can track progress, identify gaps, and provide valuable data for future interventions. Comparative studies with other countries will offer insights into best practices and innovative solutions that can be adapted to the Malaysian context (Dutkiewicz & Ellis, 2018). By leveraging technology, implementing strategic policy reforms, enhancing education and capacity-building efforts, fostering supportive networks, and committing to ongoing research, Malaysia can ensure that women entrepreneurs thrive and lead in the business sector.

Such comprehensive efforts will not only address existing gender disparities but also drive Malaysia's economic growth, innovation, and social equity. As the country continues to prioritise gender diversity in leadership, it has the opportunity to establish itself as a leader in inclusive entrepreneurship and business innovation, contributing to a more sustainable and equitable economy for all.

5.5 Conclusion

In conclusion, the findings of this study reveal the complex and dynamic evolution of women's participation in the Malaysian business sector, shaped by historical, policy-driven, and societal shifts. From the early post-independence era, when women primarily engaged in informal and family-based businesses, to the present, where they have increasingly assumed leadership roles and ventured into diverse industries, the progress is undeniable. Key milestones, such as the establishment of the New Economic Policy (NEP) in 1971, which emphasised inclusive economic growth, and the National Policy on Women (NPW) in 1989, marked pivotal moments in promoting gender equality and empowering women to participate more actively in economic activities. These interventions provided a foundation for subsequent initiatives, including the 11th Malaysia Plan (2016–2020), which set measurable targets to enhance female labour force participation and promote gender diversity in leadership roles.

A closer analysis of trends shows that women's labour force participation increased from 44.5% in 1990 to 55.3% by 2022, reflecting the impact of policies aimed at addressing structural barriers and cultural attitudes (Bangun et al., 2024). Furthermore, women-owned enterprises have grown significantly, supported by government-backed programs such as TEKUN Nasional and the Women Entrepreneur Financing Scheme, which have improved access to finance and mentorship. Despite these achievements, challenges persist, including deeply ingrained societal norms, limited representation in senior leadership, and unequal access to resources and opportunities (Selamat & Endut, 2020).

Critically, the historical trajectory of women's participation in business reveals that policy interventions alone are insufficient without corresponding societal changes. For instance, while the introduction of gender quotas in 2011 aimed to increase women's representation on corporate boards, the uptake remains below targets, with women comprising only 29.7% of decision-making positions as of 2023 (Halimah et al., 2019). This underscores the need for a more comprehensive approach that combines policy enforcement with cultural shifts to dismantle patriarchal mindsets that continue to hinder gender inclusivity.

Reflecting on these historical insights, the study emphasises that leveraging women's contributions is both a strategic and moral imperative for Malaysia's economic future. The achievements of Malaysian women in business demonstrate resilience and innovation, but the persistence of structural barriers highlights the necessity for sustained and multi-faceted efforts. Policymakers must strengthen and enforce gender equality legislation, ensure equal access to entrepreneurial resources, and address systemic biases that limit women's career advancement. Equally, educational institutions and businesses must foster leadership programs that equip women with the skills to navigate emerging sectors, particularly those driven by digital transformation (Md Faudzi et al., 2024).

Looking ahead, the findings suggest critical implications for policy and practice. First, understanding the historical context provides valuable lessons for crafting future policies that are not only inclusive but also adaptable to changing economic and technological landscapes. Second, the intersection of global economic shifts and digital innovation offers unprecedented opportunities for women entrepreneurs, which can be harnessed through targeted initiatives that focus on digital literacy and inclusion (Cheong, 2023). Lastly, continued research into the evolving role of women in business will be crucial for identifying and addressing emerging challenges, such as the impact of automation on traditionally female-dominated industries.

In sum, this study underscores the interconnectedness of Malaysia's historical, current, and future trajectories in advancing women's participation in business. The resilience and determination of Malaysian women serve as a beacon for addressing present-day challenges and shaping a more inclusive and equitable business environment. By aligning efforts across government, private sector, and civil society,

Malaysia can fully unlock the potential of its women entrepreneurs, contributing to national economic goals, including the realisation of the National Entrepreneurship Policy (DKN) by 2030. Achieving this vision requires collective commitment to dismantling barriers and building a business ecosystem that values and supports women's contributions. In doing so, Malaysia will not only foster greater gender parity but also set a benchmark for inclusive economic growth in the ASEAN region and beyond.

Highlights

- Digital transformation, primarily e-commerce, fintech, and remote working, will redefine opportunities for women entrepreneurs across Asia, enabling new avenues, global market access, and scalability.
- Policy reforms are needed to adopt adaptive policies, such as enforced gender quotas, pay equity laws, and flexible work models, to close leadership gaps.
- Expanding digital literacy, STEM, and sustainability training will prepare women for emerging industries like green tech and ESG-aligned businesses.
- Cross-border partnerships and knowledge-sharing can accelerate gender parity, leveraging best practices from leaders like Singapore and Thailand.
- Strengthening access to capital, mentorship, and networks will empower women entrepreneurs, especially in rural and marginalized communities.
- Shifting societal norms and challenging patriarchal mindsets are critical for fostering inclusive business environments.
- Women-led businesses will drive Asia's future economy through innovation, ethical practices, and resilience in the face of global challenges.
- Asia has the potential to set benchmarks for gender-inclusive entrepreneurship, shaping a more equitable and dynamic business landscape.

References

Ab Halim, F. (2023). Empowerment of women's leadership through various technical and vocational education and training (TVET) programmes. *Journal of TVET and Technology Review, 1*(1), 11–21.

Abdullah, S. N. (2014). The causes of gender diversity in Malaysian large firms. *Journal of Management & Governance, 18*, 1137–1159.

Abduvakhidov, A. M., Mannapova, E. T., & Akhmetshin, E. M. (2021). Digital development of education and universities: Global challenges of the digital economy. *International Journal of Instruction, 14*(1), 743–760.

Adapa, S., & Sheridan, A. (2021). A case of multiple oppressions: Women's career opportunities in Malaysian SME accounting firms. *The International Journal of Human Resource Management, 32*(11), 2416–2442.

Ahmad, S. H., Subramaniam, G., & Nasir, S. (2020). Women in management at the port sector of maritime industry in Malaysia—Is there a gender imbalance. *Academy of Strategic Management Journal, 19*(3), 1–7.

Ariffin, A. S., Baqutayan, S. M. S., & Mahdzir, A. M. (2020). Enhancing women entrepreneurship development framework: Policy & institution gap and challenges in the case of Malaysia. *Journal of Science, Technology and Innovation Policy, 6*(2), 22–33.

Bangun, W., Anatan, L., & Calista, C. (2024). Effort to gender equality in Indonesia, compared on ASEAN. *Journal of Infrastructure, Policy and Development, 8*(8), 5810.

Baqutayan, S. M. S. (2020). Women empowerment is needed to boost gender equality in Malaysia. *Current Research Journal of Social Sciences and Humanities, 3*, 96.

Bogren, M., Von Friedrichs, Y., Rennemo, Ø., & Widding, Ø. (2013). Networking women entrepreneurs: Fruitful for business growth? *International Journal of Gender and Entrepreneurship, 5*(1), 60–77.

Bullough, A., Guelich, U., Manolova, T. S., & Schjoedt, L. (2022). Women's entrepreneurship and culture: Gender role expectations and identities, societal culture, and the entrepreneurial environment. *Small Business Economics, 58*(2), 985–996.

Cheong, C. W. (2023). Political regimes and its influence on the diversity–performance relationship in Malaysia. *Asia-Pacific Journal of Business Administration, 15*(5), 746–771.

Chern, L. T. (2024). Malaysia falls 12 places in Global Gender Gap Report 2024. *The Star*.

Chipfunde, D., Yahaya, S. N., & Othman, N. A. (2021). The determinants influencing the performance of women entrepreneurs in Malaysia: A conceptual framework. *Studies of Applied Economics, 39*(4).

DOSM. (2022). Department of Statistics Malaysia.

Dutkiewicz, M., & Ellis, A. (2018). Putting Women's Economic Empowerment in the Asia Pacific at the Core of the G20.

Elias, J. (2015). Civil society and the gender politics of economic competitiveness in Malaysia. *Globalizations, 12*(3), 347–364.

Faudzi, M. S. M., Bakar, L. J. A., & Ahmad, S. (2024). Breaking barriers: Investigating technology adoption in micro, small, and medium enterprises (MSMEs) among low-income women entrepreneurs in Malaysia. *PaperASIA, 40*(5b), 126–135.

Ghazali, N., Yusoff, W. F. W., & Adnan, R. (2020). A framework for innovation management practices among women entrepreneurs of small medium enterprise (SMES) in Malaysia. In *30th International Business Information Management Association Conference-Vision* (pp. 1959–1974).

Halimah, S., Subramaniam, G., & Nasir, S. (2019). The port sector in Malaysia—Does gender diversity exist at the management level? *International Journal of Supply Chain Management, 8*(6), 172–180.

Hassan, Z., Subramaniam, C., Zin, M. L. M., Shamsudin, F. M., & Ramalu, S. S. (2019). The connection between safety compliance behaviour, safety communication and safety standard and procedure: An investigation among workers in Malaysian SME'S. *Academy of Entrepreneurship Journal, 25* (Special Issue 2), 1–11.

Hussin, S., Yusof, N., Hashim, H., Mohamad Ali, N., Alias, F., Abdullah, M. N., & Awang, M. N. (2020). Penglibatan Wanita Sebagai Penentu Dasar Dari Sudut Pandang Al Siyä€ Sah Al-Syarciyyah: Kajian Kes Sektor Perbankan Islam Di Malaysia: Women involvement in decision-making from the viewpoint of Al Siyä€ Sah Al-Syarciyyah: A case study on the Islamic banking sector in Malaysia. *Malaysian Journal of Syariah and Law, 8*(1), 69–80.

Kuppuswami, D., & Ferreira, F. (2022). Gender equality and women's empowerment capacity building of organisations and individuals. *Journal of Learning for Development, 9*(3), 394–419.

Kyrgidou, L., Mylonas, N., Petridou, E., & Vacharoglou, E. (2021). Entrepreneurs' competencies and networking as determinants of women-owned ventures success in post-economic crisis era in Greece. *Journal of Research in Marketing and Entrepreneurship, 23*(2), 211–234.

Maemunah, S., Yuliantini, & Sholihah, S. A. (2023). Women in logistics and transport: Evolution of women leaders and future research trends. In *Women in aviation: Management, talent and empowerment during crisis era* (pp. 29–47). Springer Nature.
Mahajan, R., & Bandyopadhyay, K. R. (2021). Women entrepreneurship and sustainable development: Select case studies from the sustainable energy sector. *Journal of Enterprising Communities: People and Places in the Global Economy, 15*(1), 42–75.
Manello, A., Cisi, M., Devicienti, F., & Vannoni, D. (2020). Networking: A business for women. *Small Business Economics, 55*, 329–348.
Masrom, M., & Ismail, Z. (2010). Women access to computers and internet: A Malaysian perspective. In *Women in engineering, science and technology: Education and career challenges*. (pp. 211–231). IGI Global.
Mohamed Shaffril, H. A., Samsuddin, S. F., & Abu Samah, A. (2021). The ABC of systematic literature review: The basic methodological guidance for beginners. *Quality and Quantity, 55*, 1319–1346.
Murugiah, S. (2022). Malaysia ranked 103 on global gender gap index 2022. The Edge Malaysia.
Nambisan, S. (2017). Digital entrepreneurship: Toward a digital technology perspective of entrepreneurship. *Entrepreneurship Theory and Practice, 41*(6), 1029–1055.
Noor, S., Isa, F. M., & Nor, L. M. (2021). Women empowerment through women entrepreneurship: A comparison between women entrepreneurs and fulltime housewife in Pakistan. *Interdisciplinary Journal of Management Studies (Formerly Known as Iranian Journal of Management Studies), 14*(2), 347–363.
Paul, J., Lim, W. M., O'Cass, A., Hao, A. W., & Bresciani, S. (2021). Scientific procedures and rationales for systematic literature reviews (SPAR-4-SLR). *International Journal of Consumer Studies, 45*(4), O1–O16.
Rashid, S., & Ratten, V. (2020). A systematic literature review on women entrepreneurship in emerging economies while reflecting specifically on SAARC countries. *Entrepreneurship and organizational change: Managing innovation and creative capabilities* (pp. 37–88).
Sakai, M. (2022). Women and Entrepreneurship in Muslim countries. *Women Entrepreneurs and Business Empowerment in Muslim Countries* (pp. 45–85). Springer International Publishing.
Schwak, J. (2021). *Gender politics and the pursuit of competitiveness in Malaysia. Women on board: Juanita Elias*. Routledge.
Selamat, N. H., & Endut, N. (2020). Bargaining with patriarchy and entrepreneurship: Narratives of Malay Muslim women entrepreneurs in Malaysia. *Kajian Malaysia: Journal of Malaysian Studies, 38*(1), 11–31.
Sharafizad, J. (2011). Determinants of business networking behaviour of women in small businesses. *Small Enterprise Research, 18*(2), 158–160.
Ughetto, E., Rossi, M., Audretsch, D., & Lehmann, E. E. (2020). Female entrepreneurship in the digital era. *Small Business Economics, 55*, 305–312.
World Bank. (2020). The World Bank in Malaysia.
Zaimah, R., Sarmila, M. S., Selvadurai, S., Lyndon, N., Er, A. C., & Norizam Jamian, M. (2013). The history and current status of dual-career families in Malaysia. *Asian Social Science, 9*(6), 16.

Hemavathi Gunasegaran is a lecturer at the Centre for Continuing Education, Universiti Malaya (Malaysia), specialising in business and management. Her research focuses on women's empowerment and entrepreneurship, with a strong emphasis on housewives. Her Ph.D. thesis examines key factors driving housewives toward entrepreneurship, reflecting her commitment to fostering economic independence and gender equality. Through award-winning research and innovation, she strives to create meaningful societal change, aligning with the SDG goal of eliminating gender inequality.

Part II
Sustainable Business Practices and Finance

Chapter 6
Sustainable Finance Initiatives Disclosures by Selected Banks in Malaysia

Clusters of Future Studies: Past and Futures; Environmental Futures; Sustainable Development

Corina Joseph, Saifulrizan Norizan, Mariam Rahmat, and Tina Stephen Enggong

Abstract This empirical study aims to examine the extent of sustainable finance initiative disclosures in nine Malaysian banks and financial institutions' sustainability reports or integrated reports in 2022 via content analysis. The items disclosed by more than 70% of the bank sustainability or integrated reports include: (1) Strategy to achieve net zero carbon; (2) Community empowerment programme; (3) Principles of reporting—sustainability context; (4) Principles of reporting—stakeholders inclusiveness; (5) Principles of reporting—materiality; (6) Concern by top management; (7) Green financing; (8) Financial inclusion; (9) Task Force on Climate-related Financial Disclosures (TCFD) recommendations; (10) Maqasid Shariah; (11) Climate risk/sustainability risk; (12) UN SDG; (13) Partnership; (14) Environmental well-being; and (15) Strong governance. All these reported initiatives indicated banks' commitment to the ASEAN taxonomies blueprint. The findings from this research contribute to the broader conversation on sustainable finance and provide a roadmap for policymakers, financial institutions, and investors, facilitating the acceleration of Malaysia's transition towards a greener, more resilient financial sector in the evolving landscape of the Future of Asia.

C. Joseph (✉) · S. Norizan · M. Rahmat · T. S. Enggong
Faculty of Accountancy, Universiti Teknologi MARA, Cawangan Sarawak, Shah Alam, Malaysia
e-mail: corina@uitm.edu.my

S. Norizan
e-mail: saifulrizan@uitm.edu.my

M. Rahmat
e-mail: mariamr@uitm.edu.my

T. S. Enggong
e-mail: stenia@uitm.edu.my

M. S. Balaji et al. (eds.), *Future of Business in Asia*,
https://doi.org/10.1007/978-3-031-95940-0_6

Keywords Sustainable finance initiative · Disclosures · Malaysian banks and financial institutions · Sustainability · Green finance · Climate change

6.1 Introduction

There are various quirks with the climate change issue. All sectors of the economy that are dependent on climate change, including agriculture, water, the coast, infrastructure, and health, are expected to suffer both nationally and globally (Ibrahim et al., 2016). This will have a significant effect on the global and intergenerational levels concerning the greater social goals of sustainable development.

Climate change provides numerous significant difficulties to society and the economy, particularly the financial sector. To address these issues, a deeper knowledge of the features of climate change-related risks and the reforms required to create a carbon–neutral economy is required. Risks associated with climate change have become more significant in the financial sector in recent years, particularly with the 2015 signing of the Paris Agreement and the UN 2030 Agenda for Sustainable Development (González, 2021). The economic development of ASEAN nations varies greatly, which has an impact on their ability to carry out sustainable financing initiatives. In comparison to the less developed economies in the region, countries such as Singapore and Malaysia could have more advanced frameworks. The importance of the financial sector in allocating funds towards the pressing sustainability challenge is acknowledged by the ASEAN Taxonomy Board. To support the speedy and just transition to net zero [for ASEAN], there is a need to significantly boost access to climate finance as well as knowledge and technology transfer on key priority areas, such as adaptation and mitigation measures, disaster risk reduction, and clean energy transition, as stated by the ASEAN Secretary-General in the first ASEAN State of Climate Change Report. In Malaysia, Bank Negara Malaysia (BNM) and the Securities Commission (SC) are involved in the ASEAN Taxonomy Board, which released its first version of the ASEAN Taxonomy on Sustainable Finance in November 2021. Bursa Malaysia announced improved sustainability reporting guidelines for publicly traded companies to improve sustainability practices and disclosures about the stock market (IPSF, 2022). For this reason, the paper aims to examine the extent of sustainable finance initiative disclosures in selected Malaysian sustainability reports or integrated reports banks and financial institutions in 2022.

Sustainable finance is significant because it fosters sustainable development by allocating funds to projects that benefit the environment and society. According to González (2021), the European Commission defines sustainable finance as the process of considering environmental and social factors when making investment decisions, resulting in higher and longer-term investments in sustainable operations. Sustainable finance reduces financial risks and boosts long-term profitability while contributing to the attainment of the United Nations' Sustainable Development Goals (SDGs), promoting economic growth.

The emergence of sustainable finance may be traced back to numerous historical milestones, demonstrating a tendency toward the inclusion of governance, social, and environmental factors in investment decisions. Overall, the development of sustainable finance is a reflection of increased environmental and societal consciousness as well as an understanding of the connections between social responsibility, environmental stewardship, and financial performance (Fathihani & Saputra, 2022).

Malaysia wants to become a net-zero country as soon as possible, by 2050 (IPSF, 2022). Malaysia updated its Nationally Determined Contribution (NDC) objectives in 2021, pledging to unconditionally lower the GDP's carbon intensity across the economy by 45% by 2030. Together with these promises, the government is fast-tracking its shift to a low-carbon economy by implementing several policies and strategic plans, such as aligning the national budget of Malaysia with the Sustainable Development Goals (IPSF, 2022). Regarding banking, the strategies outlined in Bank Negara Malaysia's (BNM) Financial Sector Blueprint 2022–2026 facilitate a fair and orderly transition of the economy.

6.2 Literature Review

6.2.1 Importance of Sustainable Finance Initiatives

In recent years, climate change risks have gained prominence in the financial sector, particularly after the Paris Agreement[1] and the United Nations 2030 Agenda for Sustainable Development were signed in 2015 (Gonzalez, 2021). The international financial system will play an important role in channelling financing into the investment. Substantial changes will be required to develop sustainable financing that is sufficiently standardised and transparent to ensure the efficient allocation of funds to sustainable activities (Gonzalez, 2021).

Given its significant impact on international economic activity, the banking sector bears a significant portion of the responsibility for the global warming crisis, and sustainable finance has surfaced as the solution to these problems. Banks have started to recognise that there are sustainability risks as a result of pressure from investors, customers, and regulators. Investors are increasingly looking for indications that corporations are assessing environmental risks and developing measures to solve them (Fong & Aquirre, 2023) and have the power to influence the firms in which they invest (Schoenmaker & Schramade, 2018). Companies that take early steps to safeguard the environment may easily avoid the expense of having to repair damages as they arise, shielding them from unanticipated financial losses (Fong & Aquirre, 2023).

[1] It is a legally binding agreement that was adopted by 196 countries at the 2015 Paris climate change conference (COP 21). The agreement encompasses climate change mitigation, adaptation, and finance.

6.2.2 Sustainable Finance Initiatives Disclosure in Annual Reports

According to Gai et al. (2023), regulations on sustainable finance necessitate financial institutions to increase their efforts in disclosing information on how they integrate environmental, social, and governance (ESG) factors into their operational procedures.

A study by BNP Paribas (2023) found that 420 asset owners and managers, hedge funds, and private equity companies from 15 nations in Europe (50%), Asia Pacific (28.6%), and North America (21.4%) of institutional investors are allocating capital to ventures that will have verifiable good outcomes in addition to financial returns. Although data restrictions are the most significant impediment to ESG integration, particularly in addressing financial risks presented by climate change, institutional investors are adopting ESG, particularly in portfolio management and investment choices (BNP Paribas, 2023). Overall, environmental disclosure has significantly improved in recent years, and the majority of members of the International Platform on Sustainable Finance (IPSF) have already imposed required regulatory obligations in this field. According to Setyowati (2020), financial institutions have developed action plans for sustainable financing and submitted yearly sustainability reports to the financial authority, demonstrating a high level of procedural compliance. Sardianou et al. (2021) also highlighted the relatively low alignment of materiality issues with SDGs oriented towards environmental matters in European banking institutions. The main European banks contribute very little overall to the SDGs, and the contributions made by each bank to the majority of the specific SDG targets continue to be notably uneven (Avrampou et al., 2019). Findings by Gallego-Sosa et al. (2021) show that the majority of the examined banks aim to achieve at least one of the Sustainable Development Goals (SDGs) and that the banks with the highest levels of commitment to the 2030 Agenda's Goals 11 (Sustainable Cities and Communities) and 13 (Climate Action) have more gender diversity on their boards of directors.

Bose et al. (2018) revealed that the publication of green banking regulatory guidelines in 2011 by the Central Bank of Bangladesh had a favourable impact on the amount of information related to green banking. Over time, the banking industry's green banking disclosure policies have come together and established a standard procedure. The findings by Cosma et al. (2020) indicated that banks' contributions to the SDGs appear to differ, depending on their place of origin, legal system, and adoption of an integrated report. On the other hand, it seems that a bank's business plan or its listing on a stock exchange does not serve as a distinguishing factor regarding its dedication to the Sustainable Development Goals (SDGs).

A study on Islamic banks in Malaysia reported that despite showing some progress between 2011 and 2020, their overall performance in corporate sustainability reporting remains low. It appears that Islamic banks report differently than other banks, with the majority of them performing better when it comes to SDG disclosure rankings. Jan et al. (2023) revealed that of all the Sustainable Development Goals (SDGs) in Islamic banks, SDG-1, SDG-11, SDG-17, and SDG-9 appeared to have received the highest transparency ratings. However, SDG-16 received the lowest score for disclosure, indicating that the Islamic banks included in the study did not adequately share information regarding their approach to addressing sustainability issues. Gunawan et al. (2022) explored sustainability performance and green banking practices through an analysis of disclosures in sustainability reports. Their findings suggested that Indonesian banks prioritize economic metrics over social and environmental indicators.

6.3 Methodology

A content analysis of nine banks' sustainability reports or integrated reports for the year 2022 has been carried out to achieve the objective of this paper, i.e., to analyse sustainable finance initiative information. The samples of commercial banks and international banks were strategic and aligned with the research objectives. The aim was not to provide a comprehensive view of the entire national landscape of sustainable finance but rather to analyse sustainable finance information disclosure practices among selected banks. The selection of banks for this study was driven by the availability of data. In our study, we analysed textual content on banks' sustainability reports or integrated reports to assess the presence and depth of information related to sustainable finance initiatives. The content analysis was performed in December 2023.

The entire sustainability reports or integrated reports relating to sustainable finance initiatives have been explored, read (several times), and reviewed. The information collection process involved a systematic and structured approach to extract relevant data from banks' sustainability or integrated reports. In this study, the researchers drew upon the work of Joseph et al. (2023) and adapted their principles to enhance methodological rigor and uphold the reliability of the findings. The process is summarised as follows:

1. Defining Data Categories: Before data collection, specific categories of information relevant to this study were identified, including but not limited to:

    - Environmental Management
    - Social
    - Governance
    - Sustainable Finance
    - Sustainability Initiatives
    - Principles of Sustainability Reporting

    An extensive literature review was conducted to identify common themes, best practices, and key elements of sustainable finance initiative disclosure. This literature review served as the foundational step in item selection, identifying established areas of interest within the field of sustainable finance.

2. Keyword Search: Within the identified reports sections, keyword searches using terms such as *sustainable finance*, *climate change*, *climate risk*, and *green financing* were conducted. In this way, relevant content within the relevant banks' reports could be traced efficiently.
3. Data Extraction: For each bank, data based on the predefined categories were systematically extracted. This included recording the presence or absence of specific information and details about sustainable finance initiatives. Initially, the items captured were 54 items. Further refinement of items is made to avoid redundancies. The final sustainable finance initiative checklist consists of 40 items.

The sustainable finance initiatives disclosure items under six categories are as follows:

(a) Environment management (10 items)
(b) Social (4 items)
(c) Governance (4 items)
(d) Principles of reporting (3 items)
(e) Sustainable finance (5 items)
(f) Sustainability initiatives (14 items)

The existence of sustainable finance initiatives information was awarded '1' and '0' if otherwise. To improve the reliability of findings, inter-rater reliability was performed. It was conducted to evaluate the conformity among researchers in their content analysis of sustainability or integrated reports. This procedure was essential to ensure uniformity in the analysis.

6.4 Findings

Out of nine banks under study, five are sustainability reports, two integrated reports, a combination of sustainability and integrity reports, and one specific Task Force on Climate-related Financial Disclosures report. The disclosure of sustainable finance initiatives is presented in Table 6.1.

All banks disclosed sustainability initiatives related to environmental well-being and strong governance. This finding is in line with Bose et al. (2018), who discovered that the degree of green banking disclosure is positively impacted by corporate governance systems (e.g., board size and institutional ownership). The example of disclosure of strong governance is included in the Bank Islam sustainability report:

> *We have launched five sustainability commitments, endorsed governance oversight on sustainability, and established a dedicated Sustainability Project Working Team (SPWT) – Chairman*

This indicates the importance of sustainability governance in executing sustainable finance initiatives. The items disclosed by more than 70% of the bank sustainability or integrated reports include: (1) Strategy to achieve net zero carbon; (2)

Table 6.1 Disclosure of sustainable finance initiative information in selected Malaysian financial institutions

		AM BANK	BIMB	CIMB	HLB	Maybank	MBSB	RHB	PUBLIC BANK	HSBC	
A: Environment management											
1	Sustainability initiative—environmental well-being	√	√	√	√	√	√	√	√	√	100%
2	ESG Risk Assessments (ESGRA) for financing to include climate change Considerations	√	√								22%
3	Empowering our customers to achieve decarbonisation	√			√	√		√	√		56%
4	Strategy to achieve zero carbon	√	√	√	√	√	√	√		√	89%
5	Sustainability highlight—reduction in carbon emission	√		√	√			√	√		56%
6	Group environmental policy	√			√				√		33%
7	Environmental waste management practice guide	√			√	√	√		√		56%
8	Group energy policy	√			√	√	√				44%
9	Energy management practice guide	√							√		22%
10	Water management					√	√				22%

(continued)

Table 6.1 (continued)

		AM BANK	BIMB	CIMB	HLB	Maybank	MBSB	RHB	PUBLIC BANK	HSBC	
B: Social											
1	Sustainability initiative—positive social impact	√		√	√	√	√				56%
2	Community empowerment programme	√	√	√	√	√	√	√	√		89%
3	Sustainable work culture	√	√	√		√					44%
4	Community empowerment code of conduct—positive social impact	√		√	√	√					44%
C: Governance											
1	Sustainability initiative – Strong governance	√	√	√	√	√	√	√	√	√	100%
2	Organisation structure – Board of directors	√	√			√			√		48%
3	Organisation structure – Group sustainability Council	√	√	√					√		48%
4	Organisation structure – Group sustainability Department	√	√		√	√		√			56%
D: Principles of reporting											
1	Principles of reporting – sustainability context	√	√	√	√	√	√	√	√		89%

(continued)

Table 6.1 (continued)

		AM BANK	BIMB	CIMB	HLB	Maybank	MBSB	RHB	PUBLIC BANK	HSBC	
2	Principles of reporting – stakeholders' inclusiveness	√	√	√	√	√	√	√	√		89%
3	Principles of reporting – materiality	√	√	√	√	√	√	√	√		89%
E: Sustainable Finance											
1	Concern by top management	√	√	√	√	√	√	√			78%
2	Green financing		√	√	√	√	√	√	√		78%
3	Fully implemented the BNM climate change and principles-based taxonomy (CCPT)		√	√		√			√		44%
4	Financial inclusion	√	√	√	√	√	√	√	√		89%
5	Task force on climate-related Financial disclosures (TCFD) recommendations		√	√	√	√	√	√	√		78%
F: Sustainability initiatives											
1	Business highlights – sustainability journey	√		√	√	√	√	√	√		78%
2	Sustainability highlight – contribution to SMEs via financing	√	√	√		√		√			56%

(continued)

Table 6.1 (continued)

		AM BANK	BIMB	CIMB	HLB	Maybank	MBSB	RHB	PUBLIC BANK	HSBC	
3	Sustainability highlight – contribution to local supplier	√		√	√		√		√		56%
4	Sustainability highlight – Zakat contributions	√	√								22%
5	Group sustainability framework	√	√	√		√		√		√	67%
6	Sustainability statement	√	√	√	√	√			√		67%
7	Sustainability themes	√			√				√		33%
8	Maqasid Shariah	√	√	√	√	√	√	√		√	89%
9	Climate Risk/Sustainability risk	√	√	√	√	√		√	√		78%
10	Award	√	√	√		√			√		56%
11	ESG Risk Assessment (ESG-RA) Guideline	√		√	√	√		√		√	67%
12	Sustainability performance	√	√	√	√	√	√	√	√		89%
13	UN SDG	√	√	√	√	√	√	√	√		89%
14	Partnership			√	√	√	√	√			56%
		35/40 = 87.5%	26/40 = 65%	29/40 = 72.5%	29/40 = 72.5%	31/40 = 77.5%	21/40 = 52.5%	23/40 = 57.5%	25/60 = 62.5%	6/40 = 15%	

Community empowerment programme; (3) Principles of reporting—Sustainability context; (4) Principles of reporting—Stakeholders inclusiveness; (5) Principles of reporting—Materiality; (6) Concern by top management; (7) Green financing; (8) Financial inclusion; (9) Task Force on Climate-related Financial Disclosures (TCFD) Recommendations; (10) Maqasid Shariah; (11) Climate risk/Sustainability risk; (12) UN SDG; and (13) Partnership.

One example of the principle of reporting—stakeholders' inclusiveness is as follows:

Our sustainability agenda considers balancing our diverse stakeholders' near- and long-term interests. We take into account various stakeholders, including civil society, to better serve their needs. By establishing meaningful and relevant stakeholder relationships, we create value for both our business and non-business stakeholders, thus enabling a more sustainable future for everyone (AM Bank sustainability report).

This finding is consistent with Kuhn (2022), who found that several initiatives from various stakeholder groups, such as civil society organisations, have aided in expanding and extending discourses, which has aided in the mainstreaming of sustainable finance in Germany.

Several banks explained well about climate risk. One of them is Bank Islam Malaysia Berhad:

Risk arising from the effect of climate change on the group's operations and to its customers. Climate risk includes transition risk that entails changes in policies, investments, and business strategy such as carbon taxation that may negatively impact segments that are carbon-intensive or unable to adapt promptly.

The top management plays an important role in implementing sustainable finance initiatives. This result is consistent with that of Ilyas et al. (2020), who found that community-based SDGs, the environment, and green supply chain management are all greatly influenced by top management support. This is illustrated by the concerns raised by the Group chairman in the CIMB sustainability report (2022).

With this in mind, we made a conscientious decision to double our sustainable finance commitment to RM60 billion by 2024, having achieved our initial RM30 billion targets ahead of schedule. This commitment is guided by CIMB's Green, Social, Sustainable Impact Products and Services (GSSIPS) Framework, our internal taxonomy developed to deliver impactful sustainable finance (CIMB sustainability report, 2022).

The example of climate-related financial disclosure was provided by RHB as follows:

Our management approach to climate change focuses on two areas:

- *Reducing our operational footprint to achieve carbon neutrality by 2030 and developing a pathway to achieve net zero by 2050, covering Scope 1, 2, and 3 emissions, including our financed emissions.*
- *Engaging with selected customer segments to better understand their decarbonisation strategy and providing financial support for their transition plan towards low carbon and sustainable practices.*

This result is consistent with the study conducted by Bose and Hossain (2021). It is important to provide regulators, legislators, and standard-setters with information about the financial disclosure practices of businesses. This information will aid in the development of an extensive disclosure policy that will influence these practices and ensure the long-term financial stability of the organisations involved.

An example of a good disclosure about green financing was provided by BIMB:

Through financing ESG-aligned sectors, we generate measurable social and environmental impact, creating more sustainable financing opportunities across Malaysia and expanding our presence in the wider green economy ecosystem.

MBBS emphasised the importance of sustainable financing as follows:

By effectively utilising our sustainable financing channels, we will be able to direct our investments and financing activities towards environmentally and socially responsible projects that prioritise long-term benefits over short-term gains.

This result is consistent with Ziemba (2023), the strategy to make the financial sector more sustainable requires banks to declare whether their funds face environmental, social, or corporate governance concerns, as well as how they invest in sustainability.

An important concept in sustainable finance is financial inclusion. An example of disclosure on financial inclusion is as follows:

Financial inclusion: Efforts to promote and facilitate financial literacy, inclusiveness, and accessibility to financial services for the underserved or unserved individuals and businesses, alleviating financial hardships and thus enhancing opportunities for all (Hong Leong Bank Sustainability Report).

This commitment is aligned with the findings by Mahmood et al. (2022), who asserted that sustainability in economic growth and social cohesion can be achieved in large part through the importance of financial inclusion in policy circles. Financing inclusion has improved because of the efforts of numerous international financial institutions and financial development organisations.

Disclosures about achieving the net zero target were provided by several banks. For example:

This year the Bank has deepened its commitment and renewed its carbon reduction target to achieve net zero emissions for Scope 1 and Scope 2 by 2030 and net zero for our overall emissions by 2050.

(Hong Leong Bank Sustainability Report)

Meanwhile, Maybank disclosed good information relating to the partnership:

Partnership for Goals Recognition In relation to UN SDG 17, this is to recognise and showcase organisations that have strived to implement sustainable change and impact via strong partnerships. Maybank was recognised for our partnerships with various organisations, including the Joint Committee on Climate Change (JC3), Partnership for Carbon Accounting Financials, WWF-Malaysia, 2° Investing Initiative, and ASEAN Capital Markets Forum (Maybank).

According to Saxena et al. (2021), achieving the UN SDGs will require robust guiding leadership that can foster strong relationships across the public, private, and non-profit sectors. This leadership must also be capable of influencing the perspectives of individuals, organisations, governments, and institutions.

Banks are committed to realizing the UN SDG as illustrated below:

Achieving one million hours per annum on sustainability and delivering one thousand significant UN SDG-related outcomes by 2025 Climate-related Opportunities in Sustainable Finance (Maybank Sustainability Report).

This commitment is in line with findings by Gallego-Sosa et al. (2021) that showed the majority of the examined banks (30 largest banks in Europe) aim to achieve at least one of the Sustainable Development Goals (SDGs).

In sum, the disclosure of sustainable finance information by banks indicates the commitment to promote sustainability practices as outlined in the ASEAN taxonomies and UN SDG 2030 agenda. Moreover, this commitment supports the speedy and just transition to net zero for the ASEAN countries. Thus, more promotion through disclosure practices would certainly benefit stakeholders who demand more transparency on how the banks invest in sustainability initiatives.

6.5 Discussion of Implications

The findings of this study have several implications for stakeholders, practitioners, and researchers. The study benefits the stakeholders in the way that it: (1) supports the course of actions that promote transparency in green reporting and sustainable finance practices by banks; (2) encourages stakeholder engagement practices to ensure that sustainability reporting deals with stakeholders' interests and information demands successfully; and (3) applies reporting standards for sustainable finance activities to improve reliability and comparison across financial institutions.

As for the practitioners, the study: (1) encompasses best practices in environmental management, social empowerment, and governance to create a competitive advantage and maintain stakeholders' trust; (2) enhances disclosure of sustainable finance initiatives, including green loans, sustainable investments, and impact measurement methodologies, to showcase dedication to sustainability; and (3) establishes governance structures to highlight openness and responsibility in sustainable finance activities.

Researchers can benefit from this study as they will be able to: (1) perform longitudinal studies to follow the changes and impact of sustainable finance initiatives in banks over time, analysing changes in disclosure practices; (2) explore the relationship between sustainable finance initiatives and financial performance metrics of banks to understand their impact on profitability, risk management, and shareholder value; and (3) conduct comparative analyses between banks, regions, or banking systems to identify best practices and areas for improvement in sustainable finance practices.

Case studies highlight Asia's proactive strategy in tackling sustainability issues through cooperation, ingenuity, and legislative changes, for example, green technology in China and sustainable agriculture in Southeast Asia. These Asian nations are contributing to global efforts towards a more resilient and sustainable future by utilizing green technologies, renewable energy sources, and sustainable agriculture methods. At the same time, they are lessening their environmental impact. These programmes provide important models that other areas can use to guide and inspire their sustainability efforts. In addition, these innovation solutions can channel capital towards environmentally and socially beneficial projects, driving positive environmental and social impact while generating financial returns.

6.6 Future Directions

Banks disclosed sustainability initiatives information related to environmental well-being and strong governance. As Asia continues to undergo fast economic development, environmental degradation and social inequality have turn out to be progressively important matters. The emphasis on sustainability initiatives by banks supports the wider regional program of transitioning to a low-carbon economy. Countries such as China, Japan, and Singapore are aggressively promoting green finance. Financial institutions are progressively anticipated to help plans that mitigate environmental risks. In addition, strong governance is crucial to improve trustworthiness, as Asia develops into a major global player in financial markets. Banks in Asia that concentrate on sustainability are expected to create a competitive edge, invite foreign investments, and support regional stability in the future.

The 70-percentage disclosure of sustainability governance information exhibits an increasing respect among Asian financial institutions of the significance of governance forms in ensuring effective implementation of sustainable finance initiatives. As the region progresses towards better integration of Environmental, Social, and Governance (ESG) principles, the disclosure of sustainability governance would be essential. Investors are gradually requiring clear, consistent, and comparable disclosures, which would influence the future of corporate reporting standards in Asia. Countries such as Malaysia, Thailand, and Indonesia are improving regulatory frameworks to uphold ESG disclosures, which would potentially become standard practice in the future.

Financing inclusion has improved due to the initiatives of several international financial institutions. Financial inclusion is a major focus for various Asian countries, in which large parts of the people continue to be unbanked. The increase of fintech, microfinance, and digital banking plans is transforming access to financial services in the region. Moving forward, financial institutions that positively incorporate financial inclusion into their sustainability strategies may potentially enjoy greater market access and greater community relationships, encouraging economic resilience and social stability.

6.7 Conclusion

This paper aims to analyse sustainable finance information on banks' sustainability and integrated reports from nine banks. These banks provide information related to sustainable finance in their sustainability or integrated reports. Disclosures of sustainable finance initiatives cover the categories: (1) Environmental management; (2) Social; (3) Governance; (4) Sustainable finance; (5) Sustainability initiatives; and (6) Principles of sustainability reporting.

This paper is not without any limitations. Firstly, the quality and availability of data in sustainability reports can vary significantly among different banks. Some banks may provide comprehensive and detailed information about their sustainable finance initiatives, while others might have limited or less transparent reporting. Incomplete or inconsistent data across reports can affect the analysis and conclusions drawn. Secondly, content analysis involves the interpretation and categorisation of textual data. Researchers' subjectivity in interpreting and categorising information from reports could introduce bias or inconsistencies in the analysis. Different interpretations of sustainability initiatives might lead to varying conclusions. Thirdly, sustainability reports may have different reporting periods or frequencies. Analysing reports from specific time frames might not capture the most recent developments or changes in sustainable finance initiatives. Additionally, initiatives and strategies might evolve, rendering older reports less relevant to current practices.

Highlights

- Contribute to a broader conversation on sustainable finance.
- Provide valuable insights into the progress made by Malaysian financial institutions.
- Offer a unique perspective on sustainable finance's role in shaping the future of the Asian economy, with Malaysia as a compelling case study.

References

Avrampou, A., Skouloudis, A., Iliopoulos, G., & Khan, N. (2019). Advancing the sustainable development goals: Evidence from leading European banks. *Sustainable Development, 27*(4), 743–757. https://doi.org/10.1002/sd.1938

BNP Paribas. (2023, September 6). Institutional investors accelerate their low-carbon transition strategies, BNP Paribas ESG Global Survey finds, Sustainability. Retrieved June 1, 2024, from https://securities.cib.bnpparibas/institutional-investors-accelerate-their-low-carbon-transition-strategies-bnp-paribas-esg-global-survey-finds/

Bose, S., & Hossain, A. (2021). An exploratory study on climate-related financial disclosures: International evidence. In *Corporate narrative reporting: Beyond the numbers* (pp.1–28). Taylor and Francis, Forthcoming. Retrieved June 1, 2024, from https://ssrn.com/abstract=3966757

Bose, S., Khan, H. Z., & Rashid, A. (2018). What drives green banking disclosure? An institutional and corporate governance perspective. *Asia Pacific Journal of Management, 35*, 501–527. https://doi.org/10.1007/s10490-017-9528-x

Cosma, S., Venturelli, A., Schwizer, P., & Boscia, V. (2020). Sustainable development and European banks: A non-financial disclosure analysis. *Sustainability, 12*, 6146. https://doi.org/10.3390/su12156146

Fathihani, F., & Saputra, J. (2022). A review of sustainable finance and financial performance literature: Mini-Review approach. *International Journal of Finance, Economics and Business, 1*(1), 1–17.

Fong, S., & Aquirre. (2023, October 23). What is sustainable finance and why is it important. Retrieved June 1, 2024, from https://www.rydoo.com/cfo-corner/sustainable/finance

Gai, L., Bellucci, M., Biggeri, M., Ferrone, L., & Lelasi, F. (2023). Manuscript full title: Banks' ESG disclosure: A new scoring model. *Finance Research Letters, 57*, Article 104199. https://doi.org/10.1016/j.frl.2023.104199

Gallego-Sosa, C., Gutiérrez-Fernández, M., Fernández-Torres, Y., & Nevado-Gil, M. T. (2021). Corporate social responsibility in the European banking sector: Commitment to the 2030 agenda and its relationship with gender diversity. *Sustainability, 13*(4), 1731. https://doi.org/10.3390/su13041731

Gonzalez, C. I. (2021, September 8). Overview of global and european institutional sustainable finance initiatives. Banco de Espana Article 30/21. Retrieved June 1, 2024, from https://ssrn.com/abstract=3937147

Gunawan, J., Permatasari, P., & Sharma, U. (2022). Exploring sustainability and green banking disclosures: A study of banking sector. *Environment, Development and Sustainability, 24*(9), 11153–11194. https://doi.org/10.1007/s10668-021-01901-3

Ibrahim, K., Shabudin, A. F. A., Chacko Koshy, K., & Asrar, G. R. (2016). A new framework for integrated climate finance and inclusive responses to sustainable development in Malaysia. *Geomatics, Natural Hazards and Risk, 7*(6), 1754–1768. https://doi.org/10.1080/19475705.2016.1155503

Ilyas, S., Hu, Z., & Wiwattanakornwong, K. (2020). Unleashing the role of top management and government support in green supply chain management and sustainable development goals. *Environmental Science and Pollution Research, 27*, 8210–8223. https://doi.org/10.1007/s11356-019-07268-3

International Platform on Sustainable Finance (IPSF) Annual report 2022. (2022). Retrieved June 1, 2024, from https://finance.ec.europa.eu/system/files/2022-11/221109-ipsf-annual-report_en.pdf

Jan, A. I., Lai, F. W., Asif, M., Akhtar, S., & Ullah, S. (2023). Embedding sustainability into bank strategy: Implications for sustainable development goals reporting. *International Journal of Sustainable Development & World Ecology, 30*(3), 229–243. https://doi.org/10.1080/13504509.2022.2134230

Joseph, C., Norizan, S. & Rahmawati. (2023). An analysis of research ethical practices information on universities' websites in developing and developed countries. *Journal of Academic Ethics*. https://doi.org/10.1007/s10805-023-09496-2

Kuhn, B. M. (2022). Sustainable finance in Germany: Mapping discourses, stakeholders, and policy initiatives. *Journal of Sustainable Finance & Investment, 12*(2), 497–524. https://doi.org/10.1080/20430795.2020.1783151

Mahmood, S., Shuhui, W., Aslam, S., & Ahmed, T. (2022). The financial inclusion development and its impacts on disposable income. *SAGE Open, 12*(2). https://doi.org/10.1177/21582440221093369

Sardianou, E., Stauropoulou, A., Evangelinos, K., & Nikolaou. (2021). A materiality analysis framework to assess sustainable development goals of banking sector through sustainability reports,

Sustainable Production and Consumption, 27,1775–1793. https://doi.org/10.1016/j.spc.2021.04.020

Saxena, A., Ramaswamy, M., Beale, J., Marciniuk, D., & Smith, P. (2021). Striving for the United Nations (UN) sustainable development goals (SDGs): What will it take? *Discover Sustainability, 2*, 1–14. https://doi.org/10.1007/s43621-021-00029-8

Schoenmaker, D. & Schramade, W. (2018). *Principles of sustainable finance*. Oxford University Press

Setyowati, A. B. (2020). Governing sustainable finance: Insights from Indonesia. *Climate Policy, 23*(1), 108–121. https://doi.org/10.1080/14693062.2020.1858741

Ziemba, M. (2023). Sustainable finance—Where we are and how we can go further. *Finanse i Prawo Finansowe, 1*(37), 123–137. https://doi.org/10.18778/2391-6478.1.37.07

Corina Joseph graduated from Curtin University, Australia, and is currently a full Professor at the Faculty of Accountancy, Universiti Teknologi MARA, Cawangan Sarawak, Malaysia. She is an Associate Member of the Malaysian Institute of Accountants and an Associate Fellow of the Accounting Research Institute, one of the Higher Institutions Centre of Excellence under the Ministry of Higher Education, Malaysia. Her research interests are in Corporate Social Responsibility (CSR) and Sustainability, Corporate Governance, Public Sector Accounting, and Financial Criminology. She is currently the Co-Chief Editor of the International Journal of Service Management and Sustainability, one of the university's homegrown journals. She also serves as the Editorial Advisory Board Member of several Scopus and international refereed journals. Corina worked as an Adjunct Professor at the Faculty of Economics and Business, Universitas Pembangunan Nasional 'Veteran', East Java, Indonesia. Corina has been selected by the university to participate in the Distinguished Scholar Program under the Standing Committee for Scientific and Technological Cooperation (COMSTECH) dedicated to the promotion and cooperation of science and technology activities among the OIC member states. Corina's expertise has been acknowledged and included in the World List of Sustainability Researchers compiled by the Inter-University Sustainable Development Research Programme (IUSDRP), Hochschule für Angewandte Wissenschaften Hamburg/Hamburg University of Applied Sciences—HAW Hamburg. Corina has contributed three chapters i.e. two from SDG 4—Quality Education" and one on SDG 12—Responsible Consumption and Production in the 'Encyclopaedia of the UN Sustainable Development Goals: Transforming the World We Want 'published by Springer and are associated with the "World Sustainability Series", consequently being a high impact, high visibility, peer-reviewed output. Corina is actively supervising both Master by Research and Doctorate students and playing a role as a mentor for local and international academicians in terms of research and publication.

Saifulrizan Norizan is a lecturer at Faculty of Accountancy, Universiti Teknologi MARA, Cawangan Sarawak, Malaysia. He graduated with a Master of Accounting from UiTM Shah Alam. His area of specialisation are financial accounting and reporting and taxation. He has more than ten years teaching experience for undergraduate accounting courses and professional courses (ACCA and ICSA). He is also an associate member of Malaysian Institute of Accountants. Apart from teaching, he is also actively engaged in doing research and writing for journal publication.

Mariam Rahmat is a Senior Lecturer at the Faculty of Accountancy, Universiti Teknologi MARA, Cawangan Sarawak, Malaysia. She holds a Bachelor of Accountancy (Hons) from Universiti Utara Malaysia and Master of Accounting from Curtin University of Technology. She is also an associate member of Malaysian Institute of Accountants. She has been an accounting lecturer for UiTM since 1992 to diploma and undergraduate students specializing in financial accounting and reporting.

Tina Stephen Enggong is an accounting lecturer at Faculty of Accountancy, Universiti Teknologi MARA, Cawangan Sarawak, Malaysia. Since 2001, she has been a permanent faculty member, teaching financial accounting, cost and management accounting, and computerised accounting at the diploma level.

Chapter 7
Towards a Sustainable Business Model in Light of Inclusiveness, Circularity, and Green Business Practices—A Conceptualisation

Clusters of Future Studies: Environmental Futures; Past and Futures

Anh Tuan Tran, Thu Ha Nguyen, and Anh Tuan Nguyen

Abstract In an era of globalisation, businesses face growing pressure to address environmental, social, and economic challenges. The United Nations Sustainable Development Goals (SDGs) offer a universal framework to guide inclusive and sustainable business practices. However, emerging business models, such as circular and inclusive approaches, remain fragmented, creating challenges for widespread adoption and standardisation. This conceptual paper examines sustainable business models by integrating economic, environmental, and social dimensions of sustainability. Focusing on Vietnam, a rapidly developing country with significant foreign direct investment (FDI) inflows, this study explores how sustainable business practices can foster long-term growth and global competitiveness. Using secondary data from company reports and national policies, as well as insights from literature on circular and inclusive business models, the research develops a synergistic framework for sustainable business models tailored to Vietnam's context. The findings highlight the importance of aligning business practices with sustainability goals and emphasise the need for empirical studies to further refine these models. This study

A. T. Tran (✉)
University of Economics and Law, Vietnam National University Ho Chi Minh City, Ho Chi Minh City, Vietnam
e-mail: tatuan@uel.edu.vn

Budapest University of Economics and Business, Budapest, Hungary

T. H. Nguyen
Faculty of Foreign Languages, Van Lang University, Ho Chi Minh City, Vietnam
e-mail: ha.nguyenthu@vlu.edu.vn

A. T. Nguyen
School of Hospitality and Tourism, Hue University, Hue, Vietnam
e-mail: Natuan@hueuni.edu.vn

M. S. Balaji et al. (eds.), *Future of Business in Asia*,
https://doi.org/10.1007/978-3-031-95940-0_7

provides a foundation for future research and practical applications in advancing sustainable business strategies in Vietnam and beyond.

Keywords Sustainable business model · Circular economy · Inclusive business · Synergy · Stakeholders management

7.1 Introduction

The stress on the planet's finite natural resources is intensifying as the global population grows exponentially and consumption rises rapidly. Increased consumption has led to an alarming rise in waste generation, contributing to global warming and widespread contamination of land, oceans, and rivers. In response, there is a global call for governments and businesses to embrace sustainable practices, particularly the circular economy. Businesses operating within the conventional linear economy continue to benefit economically, yet mounting pressures from governments, communities, and consumers highlight the need for a paradigm shift (Geissdoerfer et al., 2017). This shift necessitates rethinking business models to balance economic success with environmental and social responsibilities. Teece (2010) described the business model as the "design and architecture" for creating, capturing, and delivering value—a framework well-suited for incorporating sustainability principles. The circular economy has gained traction as a sustainable alternative, underpinned by the three pillars of sustainability—society, the environment, and the economy (Korhonen et al., 2018). These pillars provide a pathway for transitioning from a linear to a circular economy, emphasising the need to integrate circular principles into business strategies. However, adapting to a circular economy is not without challenges, as it often requires reconciling sustainability goals with existing business values and ideologies (Hofmann, 2019). This research explores the concept of a sustainable business model as a synergistic framework that combines elements of circular and inclusive business models. By engaging multiple stakeholders, including businesses, policymakers, and communities, this study develops a conceptual framework that incorporates economic, social, and environmental considerations. The findings aim to offer practical insights for fostering sustainable business practices in an era of increasing global challenges.

In the context of increasing global challenges, businesses are under mounting pressure to adopt sustainable practices that address environmental, social, and economic dimensions. The United Nations Sustainable Development Goals (SDGs) provide a framework for guiding these efforts, yet the integration of sustainable business models, particularly those emphasising circularity and inclusiveness, remains fragmented and inconsistent across different sectors and regions (Antikainen & Valkokari, 2016; Rezaee et al., 2017). This conceptual paper aims to explore the development of a sustainable business model that harmonises these dimensions, particularly within the context of Vietnam—a rapidly developing nation that is experiencing significant foreign direct investment (FDI) inflows (Rezaee, 2016).

The transition from a linear to a circular economy (CE) is critical for sustainable development, as it emphasises the need for businesses to rethink their value creation processes to minimise waste and resource consumption (Antikainen & Valkokari, 2016; Ferraz & Pyka, 2023). The circular economy (CE) is built on three pillars: economic viability, environmental protection, and social equity (Antikainen & Valkokari, 2016). However, the shift towards this model presents challenges, including the need for businesses to reconcile existing operational ideologies with new sustainability goals (Antikainen & Valkokari, 2016). The stakeholder theory plays a pivotal role in this transition, as it underscores the importance of engaging various stakeholders—including suppliers, customers, and communities—in the sustainability journey (Svensson et al., 2018). Effective stakeholder management is essential for fostering long-term business sustainability and ensuring that diverse interests are considered in decision-making processes (Amran & Ooi, 2014; Svensson et al., 2018).

In Vietnam, the integration of sustainable business practices is particularly pertinent due to the country's rapid economic growth and the influx of foreign investments that can either contribute to or detract from sustainability efforts (Rezaee, 2016). By utilising secondary data from company reports and national policies, this study aims to develop a synergistic framework that aligns business practices with sustainability goals tailored to Vietnam's unique context (Rezaee, 2016). The findings will highlight the necessity of empirical studies to refine these models further, providing a foundation for future research and practical applications in advancing sustainable business strategies (Antikainen & Valkokari, 2016; Suriyankietkaew & Avery, 2016).

The conceptualisation of a sustainable business model that incorporates circular and inclusive practices is vital for addressing the pressing challenges of our time. By engaging stakeholders and aligning business strategies with sustainability goals, companies can enhance their global competitiveness while contributing to the broader objectives of sustainable development (Sehnem et al., 2019). This research not only aims to contribute to the academic discourse on sustainable business models but also seeks to inform practical applications that can drive meaningful change in Vietnam and beyond.

7.2 Methodology

This study employs a conceptual framework approach, as outlined by Jaakkola (2020), to explore the synergistic integration of circular economy and inclusive business models into a sustainable business model. The methodology follows a transdisciplinary approach, drawing insights from diverse fields and integrating both theoretical and practical perspectives. A critical literature review forms the foundation of the study, synthesising existing research on circular economy principles and inclusive business models to identify intersections and gaps. This review is conducted systematically, adhering to established procedures for conceptual and review-based studies, ensuring a rigorous and comprehensive analysis. Relevant peer-reviewed articles, books, and reports are selected from reputable databases and journals, focusing on works listed in high-ranking sources such as ABDC and ABS.

Additionally, policy analysis is employed to provide a practical dimension, analysing frameworks and guidelines from international organisations such as the United Nations and the World Bank. These policies are examined to understand their implications for fostering sustainable business practices and to validate the conceptual framework developed in this study. The research aims to construct a robust conceptual framework for sustainable business models that encapsulates economic, environmental, and social dimensions. This framework offers a comprehensive understanding of how circular and inclusive practices can be integrated into business strategies to achieve long-term sustainability.

The integration of sustainable business practices is particularly pertinent in Vietnam due to the country's rapid economic growth and the influx of foreign investments that can either contribute to or detract from sustainability efforts. By utilising secondary data from company reports and national policies, this study aims to develop a synergistic framework that aligns business practices with sustainability goals tailored to Vietnam's unique context. The findings will highlight the necessity of empirical studies to refine these models further, providing a foundation for future research and practical applications in advancing sustainable business strategies.

7.3 Key Concepts and Issues

7.3.1 Circular Business Model (CBM)

Closed material loops and the prolonged product life cycle are marked as crucial components of the circular economy (CE) embedded in the business model (Geisendorf & Pietrulla, 2017; Tran, 2023). From the viewpoint of business model innovation, integrating CE characteristics into value creation, value proposition, and value delivery is essential for business model development (Chen, 2020; Nußholz, 2018). Velte and Steinhilper (2016) emphasised that the circular business model (CBM) contradicts the linear model, which relies on single-use products and results in the loss of value-added at the end of a product's life. In contrast, CBM optimises life cycle processes to preserve economic and environmental value, enhancing value creation and proposition (Geissdoerfer et al., 2020; Pollard et al., 2023).

Vietnam is advancing rapidly in modernisation and urbanisation, driven by heavy industrialisation. This economic growth has led to increased pollution and environmental degradation (Tran, 2023). Consequently, climate change impacts every aspect of socioeconomic development. To address this, Vietnam must strategise policies that balance economic, environmental, and social aspects of sustainable development. In 2019, the Politburo established a development policy aligned with environmental preservation and resource management principles. This initiative was followed by Vietnam's COP26 commitment to achieve net-zero CO_2 emissions by 2050, posing significant challenges for the country to sustain economic growth while mitigating environmental damage. In fact, the Vietnamese government has paved the way for the sustainable development through policy issuances since 2012 with a focus on investment orientation as showed in Fig. 7.1.

Figure 7.1 describes the process of creating a business environment in Vietnam's market. It is notable that the Vietnamese government has launched an intensive legal framework to pave the way for sustainability. In 2018, legal issuances of the government were associated with the financial means and investments to welcome foreign investment to the market and to Vietnamese companies (see Fig. 7.1). The sustainability index was linked with the guideline for disclosure on the stock market, helping large capitalised companies to enhance their innovative dynamics to cope with the environmental issues through attracting investments. These events were marked before the commitment of the Vietnamese government at the 2021 COP26. These acts forced companies operating in Vietnam to foresee the market condition is creating opportunities for those who embrace the circular business model. Among 3 elements of the business model, which are economic, social, and environmental elements, the CBM emphasises environmental practices (Nikolaou et al., 2021). The business model embodies the circular strategies that could reduce the release of waste and shift the focus to increasing product longevity. Hence, the CBM indicates a complex organisational process of closed-loop resources for the above purposes (Corvellec et al., 2021). Nußholz (2018) formed two ways of resource management which were downstream and upstream supply chains. It means obtaining secondary material sources, which might be considered as wastes or redundant sources of other companies for the product, or collecting and recycling the used products (Wells & Seitz, 2005; Witjes & Lozano, 2016).

The adoption of the Circular Economy (CE) continues to be a subject of debate. Identifying the key aspects of the circular business model (CBM) remains challenging, as businesses must balance meeting stakeholders' economic and environmental sustainability demands (Ghisellinia et al., 2016). However, existing literature provides insights into specific indicators for transitioning to the CBM, focusing on eco-friendly business operations, product lifecycle management, and maintenance (Jensen et al., 2019; Korhonen et al., 2018; Rossi et al., 2020). Businesses often proceed cautiously with CBM adoption due to potential conflicts among stakeholders, particularly when aligning economic goals with environmental objectives. While CBM promotes both monetary and non-monetary values through environmental initiatives, it is vital to ensure profitability without compromising sustainability. Hoffman et al. (2022) emphasise that achieving circularity requires harmonising economic, social, and environmental elements alongside effective stakeholder management. Thus, the integration of these elements forms the foundation of sustainability, supporting the development of CE and its alignment with national economic frameworks (Geissdoerfer et al., 2016).

The concept of CE is still evolving, particularly regarding the "3R principles"—Reduce, Reuse, and Recycle. Over time, these principles have been expanded, with "Recovery" added in the European Commission's Waste Framework Directive (2008), and the more comprehensive "9Rs framework" proposed by Potting et al.

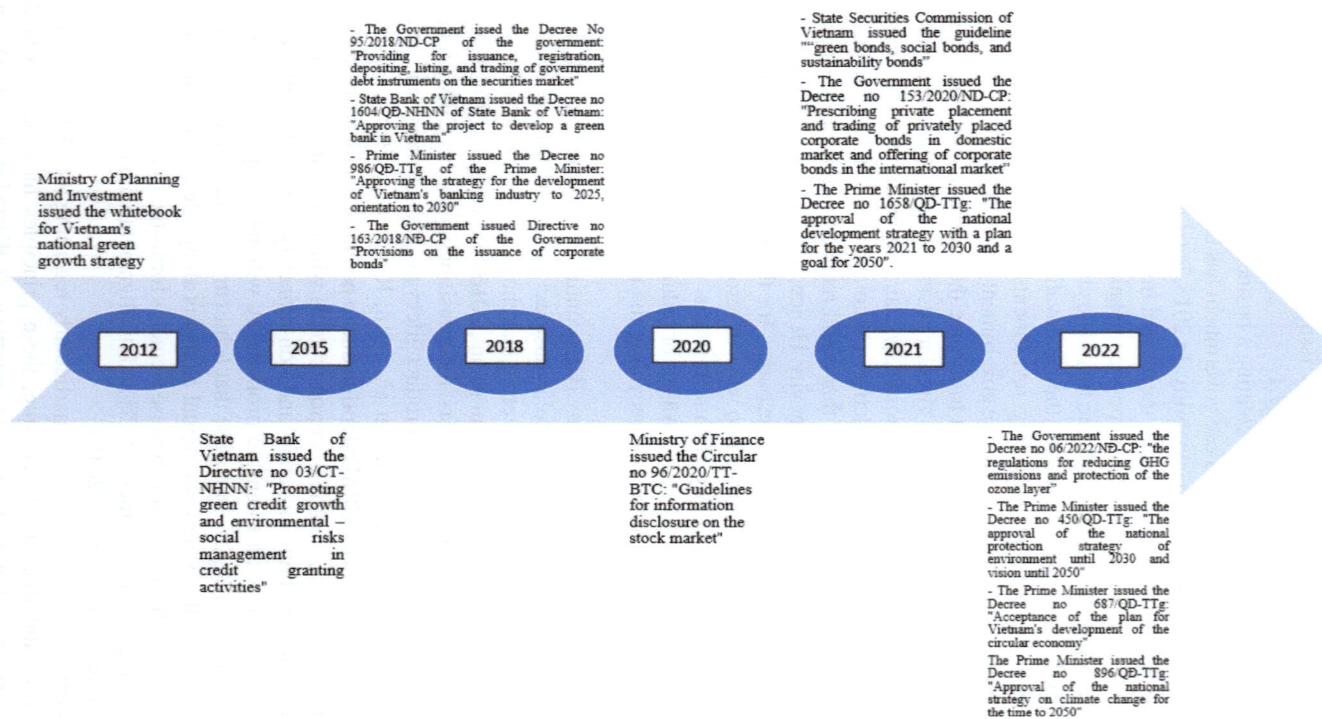

Fig. 7.1 Timeline of the sustainability-related policy issuances from 2012 to 2022. *Source* Own compilation

(2017). While these expanded frameworks provide valuable insights, the 3R principles remain more widely adopted due to their applicability in diverse national contexts and their endorsement by multiple governments (Nguyen et al., 2020). The 3R principles influence all stages of a business model, including value creation, proposition, and capture, and their application in material usage, manufacturing, recycling, and waste collection can generate economic and environmental benefits (Korhonen et al., 2018; Nußholz, 2018). Nandi et al. (2020) suggest that a successful CBM requires technological innovation, enhanced customer engagement, and sustainable revenue streams. However, integrating environmentally friendly technologies into business operations can lead to increased costs for businesses and higher prices for customers. This may also reduce financial returns for investors and shareholders. Vietnam's voluntary adoption of the Extended Producer Responsibility (EPR) policy in 2024 exemplifies its commitment to promoting environmentally friendly strategies. Nonetheless, challenges persist. Rising operational costs could lead to measures such as workforce reductions, exacerbating social inequalities in a developing economy where GDP is relatively low, and financial constraints are significant. This highlights the importance of integrating the circular business model (CBM) with the inclusive business model (IBM) to address the needs of marginalised populations while fostering sustainable development. Vietnam has also facilitated access to funding and foreign direct investment (FDI) through progressive policies, enabling businesses to adopt innovative and sustainable practices. Combining CBM with IBM ensures that economic, environmental, and social objectives are harmonised, paving the way for a more sustainable and inclusive economic framework.

7.3.2 Inclusive Business Model (IBM)

Vietnam's rapid economic growth has significantly reduced poverty levels; however, the threat of poverty resurgence persists due to economic instability and structural challenges (Schulte et al., 2023). Globalisation further complicates this scenario, requiring Vietnam to adopt innovative technologies to remain competitive. Concurrently, the global call to advance the Circular Economy (CE) to address climate change and environmental issues has intensified. Despite progress, many rural households remain at risk of falling back into poverty, highlighting the unsustainability of current poverty alleviation measures in certain regions.

The inclusive business model (IBM) has emerged as a response to the demands of stakeholders and the need for greater corporate social responsibility (CSR). Golja and Požega (2012) describe inclusive business as a model that integrates social, economic, and environmental responsibilities into business practices. Stakeholders increasingly expect companies to adopt inclusive and equitable goals that align with global market demands (Seelos & Mair, 2004). SNV & World Business Council for Sustainable Development (2008) defines inclusive businesses as those that integrate low-income populations into their value chains while balancing profit generation and social impact. Without a focus on the social dimension of business activities, inclusivity cannot be achieved.

Traditionally, inclusive business models have focused on smallholders and agricultural workers (Ros-Tonen et al., 2019). This approach ensures that low-income populations gain access to essential resources like food and employment while participating in value chains (Kelly et al., 2015; Oberlack et al., 2023). The United Nations Development Programme (UNDP, 2010) recognises IBM as a means to empower low-income individuals to engage in markets and strengthen supply chains. Similarly, Likoko and Kini (2017) emphasise that involving marginalised groups in roles such as suppliers and producers enhances both socioeconomic value and business sustainability.

IBM primarily addresses the needs of bottom-of-the-pyramid (BoP) communities, aiming to foster economic opportunities and reduce inequalities (Kolk et al., 2013). Schoneveld (2020) highlights that IBM allows businesses to engage deeply with BoP communities while aligning with the Sustainable Development Goals (SDGs). Inclusive growth, as defined by Ranieri and Ramos (2013), goes beyond poverty reduction, emphasising active participation of marginalised groups in production and business processes. This inclusive approach differs from traditional pro-poor growth strategies, which primarily target poverty and inequality reduction (Klasen, 2010). Since then, numerous international organisations have been refining the definition of inclusive business, as illustrated in Table 7.1.

Table 7.1 presents key definitions of IBM by various international organisations. For example, the UNDP, FAO, and WBCSD emphasise an outcome-oriented perspective, advocating for shared benefits between businesses and BoP communities. This approach encourages businesses to adapt strategies that promote co-value creation. Meanwhile, organisations such as the International Finance Corporation (IFC) and Asian Development Bank (ADB) adopt a more income-centric perspective, focusing on financial empowerment and market access for low-income populations. In this sense, Fig. 7.2 illustrates the dynamic relationship between firms and BoP communities, emphasising their reciprocal contributions to value chains.

According to Fig. 7.2, firms provide affordable goods, services, and employment opportunities to BoP communities, while BoP communities offer firms raw materials, productivity, and market demand. This reciprocal relationship fosters sustainable supply chains and business resilience (Golja & Požega 2012; Schoneveld, 2020). Additionally, inclusive business addresses broader social issues such as inequality and gender disparity by integrating diverse smallholders into value chains (Speelman et al., 2019; The Inclusive Business Action Network, 2021).

Vietnam's agricultural-based economy and ongoing industrialisation make IBM particularly relevant. Persistent income gaps and rural poverty demand business models that address these disparities (Nguyen et al., 2021; Phan et al., 2023). Furthermore, Vietnam's commitment to achieving net-zero carbon emissions necessitates dynamic innovation to meet CE goals while addressing social challenges (Tran et al., 2023). Businesses are under pressure to adopt inclusive and collective strategies that align with environmental and social priorities.

Table 7.1 Definition of inclusive business of international organisations

Sources	Definition	Orientation
UNDP (2013, 2016)	Inclusive business models integrate the underprivileged into a company's supply chains as workers, manufacturers, and entrepreneurs, or provide reasonable prices for goods and services that the communities' people require	Outcome
SNV and World Business Council for Sustainable Development (2008, 2011)	An inclusive business is a combination incentive of economic, social, and environmental responsibilities endeavor that incorporates low-income people into the value chain of companies. In this way, it creates a shared advantage for the particular group and the business. It aims to raise profits for the business while enhancing the lives of those with low incomes	Outcome
G20 development working group (2015)	Inclusive business is about the responsibilities of companies regarding the supply of goods, services, and opportunities for BoP's people while integrating them into the value chain as all types of stakeholders	Outcome
Kelly et al. (2015) of the FAO	IBMs support smallholders' entry into markets on the grounds that both businesses and poor farmers gain from this integration	
International Finance Corporation (2018)	IBMs integrate low-income groups into companies' core activities. By using the models, businesses enhance the ability of low-income farmers and entrepreneurs in terms of finance for the expansion of access for BoP populations	Income
Beckwith (2018) of the Asian Development Bank (ADB)	Inclusive business creates a high impact on social development by enhancing BoP's access to goods and services and providing income and opportunities for BoP's employees and employers	Income

Source Own compilation

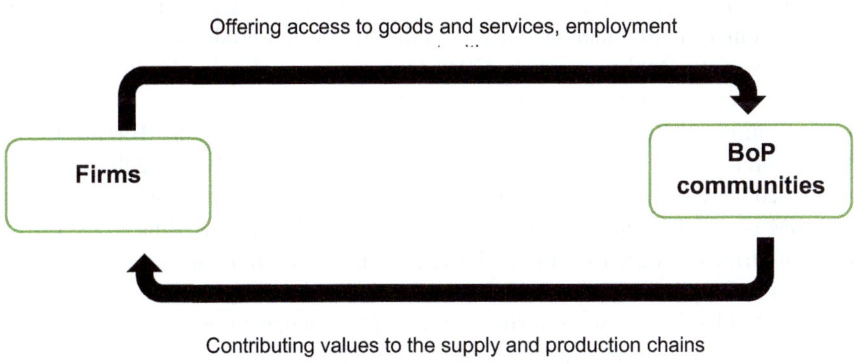

Fig. 7.2 Relationship between firms and bottom-of-the-pyramid (BoP) communities. *Source* Own compilation

The next section introduces a conceptual framework that integrates circular and inclusive business models to address Vietnam's unique challenges.

7.4 Discussion of Implications

To keep up with the global demands and challenges regarding advancing environmental protection and ecosystem, as well as also not to exclude smallholders as the country is acknowledged for its cheap labor workforce and high development rate, an integrative conceptualisation of sustainability must be pre-defined based on the circular economy and inclusive business. There is no doubt that the CBMs and the IBMs are crucial for their archetype addressing the deep integration of environmental and societal issues into companies' business models. However, IBM focuses on providing goods, services, and opportunities for people at the BoP (Bonnell & Veglio, 2011; Schoneveld, 2020) while CBM aims at reducing resource consumption, waste emission, and environmental-friendly products leading to the reduction of negative impact on the environment (Lewandowski 2016; Romero et al., 2021). The value creation of the CBM utilises the economic value that remains in items after they are used to create new offers (Linder & Williander, 2015; Pla-Julián & Guevara, 2019). In the same vein, the attributes of inclusive business and circular business have been adopted to the traditional business model (please see the Table 7.2).

Table 7.2 offered a context describing different approaches of the two business models based on the study of business canvas. While IBM focuses on the needs of stakeholders and satisfies their demands, the concentration of CBM is on the process and strategy redesign. Moreover, the propositions of IBM and CBM are different from the targets chosen for businesses. Table 7.2 demonstrates an interconnection between CBM and IBM in terms of supply and value chain while they need to integrate smallholders and new production chains. However, IBM is focusing on how to include smallholders and actors in the supply and production chain creating value for society. On the other side, CBM concentrates on the circular strategy which enhances the circularity of the product and sustainable production. A sustainable business model (SBM) is the answer to interrelationship between the IBM and CBM. Boons and Lüdeke-Freund (2012) and Hoffman et al. (2020) advocated that the SBM addressed three core issues which were social, economic, and environmental pillars.

CBM is environmental-centric while IBM is social-centric leading to the difference of strategies based on the archetype of the firms. Implementing CBM is to harmonise the pillars of sustainability and circular economy features to withstand the stakeholders' demands (Hoffman et al., 2020) when firms develop a sustainable ontology such as a sustainable production chain, reduction of waste and energy use, and sustainable material purchasing (Lewandowski, 2016). It means that in this sense, the CBM and IBM are considered as subset models to achieve sustainability. Nevertheless, the business model is put under concentration due to the value creation for society and economics and stakeholders. CBM takes the risk of the market coming from stakeholders as they could increase the price of products more than the

Table 7.2 Attributes of inclusive business model (IBM) and circular business model (CBM)

The circular business model framework proposed by Whalen (2019) based on the work of Nußholz (2018)	Inclusive business model proposed by Schoneveld (2020)	The circular business model proposed by Lewandowski (2016)
Value proposition: – "Who are the customers and what is the value to them?" – "What types of product/service(s) are the focus of this business model? " – "What circular strategies are performed?"	"An IBM's value proposition captures what value is created for what types of income-constrained groups through what types of solution offerings, as well as how a competitive advantage is derived from doing so"	"Value propositions—offered by circular products enabling product-life extension, product-service system, virtualised services, and/or collaborative consumption. Moreover, this component comprises the incentives and benefits offered to the customers for bringing back used products" (Lewandowski, 2016, p. 6)
Value creation and delivery: "How are activities, partners, channels, and resources arranged?"	"An IBM's value creation and delivery system captures how a business intends to create value. Using a theory of change, IBM articulates the mechanisms through which value will be created and monitored and the types of partnerships and value networks needed to do so effectively. This theory of change explicitly accounts for anticipatable unintended consequences arising along the path to value creation. It furthermore details how primary beneficiaries will be targeted and non-discrimination will be ensured"	"Circular value creation, pertaining to the short cycle, where products and services are maintained, repaired and adjusted, to the long cycle which extends the lifetime of existing products and processes, to cascades based on creating new combinations of resources and material components and purchasing upcycled waste streams, to pure circles in which resources and materials are 100% reused, to dematerialised services offered instead of physical products and to production on demand" (Lewandowski, 2016, p. 6)
Value capture: – "How does the firm make money? – What costs are involved?"	"An IBM's value capture system details how the business intends to capture economic value for itself from creating value for its stakeholders. Additionally, it articulates how the economic value captured will be invested in deepening and/or broadening value creation"	"Revenue streams—relying on the value propositions and comprising payments for a circular product or service, or payments for delivered availability, usage, or performance related to the product-based service offered. Revenues may also pertain to the value of resources retrieved from material loops … Cost structure—reflecting financial changes made in other components of CBM, including the value of incentives for customers. Special evaluation criteria and accounting principles must be applied to this component" (Lewandowski, 2016, p. 21)

Source Compiled by the author from Lewandowski (2016), Nußholz (2018), Schoneveld (2020), Whalen (2019) linear economic model (Mishra et al., 2018; Todeschini et al., 2017). Furthermore, CBM meets other challenges besides the market like cost increases, organisational decisions, knowledge transfers, and institutions (Vermunt et al., 2019). Boons and Lüdeke-Freund (2012)claimed that SBM created value based on the interrelationship with stakeholders by proposing 3 levels: organisational, inter-organisational, and societal. Multi-stakeholder management from reactive, proactive, and accommodated approaches is incorporated into the business model to create values for various types of stakeholders in the long-term vision (Geissdoerfer et al., 2018). Wells (2015) stated that companies enjoyed SBM by the economic profit combining the reduction of environmental and socioeconomic matters at the absolute or relative lever. Pieroni et al. (2019) backed this idea when they concluded that there was no absolute CBM or SBM which was defined by the sustainable practices and principles. Therefore, to establish a proper sustainable business, companies must incorporate the elements of CBM and IBM as crucial aspects of their business model (see Fig. 7.3).

Figure 7.3 describes that a sustainable business model must capture and integrate the core values of circular business models (CBM) and inclusive business models (IBM), which are respectively environmental-centric and social-centric. Together, these models form the foundation for achieving sustainability and aligning business operations with the 17 Sustainable Development Goals (SDGs) established by the United Nations (UN).

- CBM focuses on efficient resource utilisation, waste reduction, and life-cycle management, addressing critical environmental challenges and directly contributing to SDGs such as SDG 6 (Clean Water and Sanitation), SDG 7 (Affordable and Clean Energy), SDG 12 (Responsible Consumption and Production), and

Fig. 7.3 Components of a sustainable business model. *Source* Own compilation

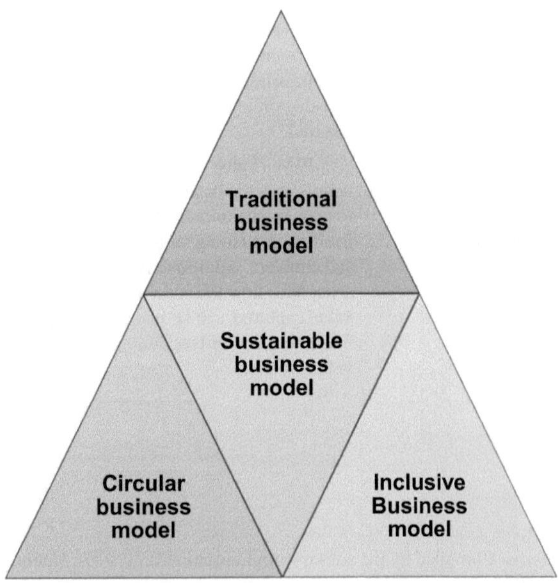

SDG 13 (Climate Action) (Butt et al., 2023; Kurnia et al., 2023). For instance, the adoption of CBM by businesses helps reduce greenhouse gas emissions and conserves finite resources, aligning with global climate goals.
- IBM emphasises equitable growth and social inclusion, addressing critical social challenges such as poverty, inequality, and lack of access to opportunities. By integrating BoP communities into value chains, IBM contributes to SDG 1 (No Poverty), SDG 4 (Quality Education), SDG 8 (Decent Work and Economic Growth), and SDG 10 (Reduced Inequalities) (Mangnus, 2019). For example, businesses using IBM models provide training and job opportunities for marginalised groups, fostering socioeconomic resilience.
- The convergence of CBM and IBM in a sustainable business model emphasises the interdependence of environmental and social sustainability as a foundation for economic sustainability. This interplay ensures businesses can contribute to SDGs while maintaining profitability and long-term viability. For example, a company adopting both CBM and IBM can reduce operational costs through resource efficiency (CBM) while fostering stronger supplier relationships and customer loyalty through inclusivity (IBM).
- With the world's resources depleting and social inequalities persisting, there is increasing urgency to integrate circularity and inclusivity into business strategies. The triangle representation in Fig. 7.3 underscores the necessity of this integration, illustrating how CBM and IBM together form a pathway for businesses to align with global sustainability objectives.

The sustainable business model is proposed by combining the attributes of ibm and cbm, as each of these models addresses several related SDGs, but not all (see Table 7.3).

Table 7.3 describes how IBM and CBM could assist any nation in achieving consecutively most SDGs. A circular economy may help with many of the urgent issues, however, the shifting process from a linear economy calls for a significant fundamental transition. Many workers and communities could suffer as a result of the first phase. Nevertheless, it is expected to result in improvement in job creation. In the case of Vietnam, The transition to a circular economy might affect the operation of businesses. Companies might need to educate new competencies for workers to increase productivity to compensate for the lower revenue.

The integration of circular business models (CBM) and Inclusive Business Models (IBM) can significantly contribute to achieving the SDGs by addressing the core challenges of sustainability and inclusivity in economic systems. Both models focus on transforming how businesses interact with resources, stakeholders, and communities.

IBM plays a critical role in addressing social challenges such as poverty, inequality, and unemployment. By embedding marginalised groups (e.g., smallholders, BoP populations) into the value chain, IBM directly targets SDG 1 (No Poverty) and SDG 8 (Decent Work). For instance, companies that include smallholder farmers as suppliers provide them with stable market access, improving food security (SDG 2) and livelihoods. Additionally, IBM fosters partnerships (SDG 17) by engaging stakeholders in shared value creation.

Table 7.3 Circular business model and inclusive business model in accordance with the framework of sustainable development goals

	Circular business model	Inclusive business model	Indirect impact
SDG 1: No poverty		X	
SDG 2: Zero hunger		X	
SDG 3: Good health & wellbeing			X
SDG 4: Quality education		X	
SDG 5: Gender equality		X	
SDG 6: Clean water and sanitation	X		
SDG 7: Affordable and clean energy	X		
SDG 8: Decent work and economic growth	X	X	
SDG 9: Industry, innovation and infrastructure	X		
SDG 10: Reduced inequalities		X	
SDG 11: Sustainable cities and communities	X	X	
SDG 12: Sustainable consumption and production	X		
SDG 13: Climate change	X		
SDG 14: Life below water		–	X
SDG 15: Life on land		–	X
SDG 16: Peace, justice, and strong institutions	–	–	X
SDG 17: Partnerships for the goals:			X

Source Own compilation

CBM emphasises environmental sustainability through the efficient use of resources, waste reduction, and lifecycle extension of products, addressing SDG 12 (Responsible Consumption and Production) and SDG 13 (Climate Action). For example, transitioning to circular operations reduces dependency on finite resources and mitigates greenhouse gas emissions. Furthermore, CBM promotes job creation in recycling, repair, and renewable energy sectors, advancing SDG 8.

The transition to a circular economy in Vietnam presents challenges for businesses, particularly in reskilling workers and restructuring operations. While the first phase of the transition might cause disruptions such as lower revenue or job losses, companies can mitigate these effects by investing in training programs to build workers' competencies. For instance, enhancing workforce skills in resource-efficient practices and green technologies aligns with SDG 4 (Quality Education) and drives productivity gains, eventually contributing to sustainable economic growth (SDG 8).

Both CBM and IBM are interconnected in addressing SDG targets. While CBM emphasises environmental benefits and long-term economic stability, IBM focuses on inclusivity and equitable growth. Together, they create a synergistic impact that strengthens businesses' contributions to sustainable development at local, national, and global levels, as illustrated in Fig. 7.4.

From Fig. 7.4, the application of an inclusive circular business model could help businesses increase productivity and inclusivity during the transition from a linear model to a circular model while avoiding laying off employees to compensate for operational costs. However, this model suggests the inclusion of poor people and

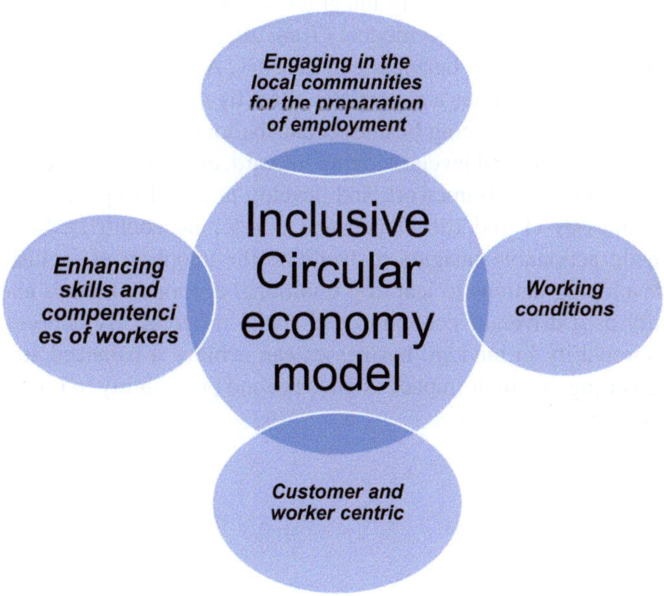

Fig. 7.4 The proposed model of the inclusive circular business model. *Source* Own compilation

BoP communities in the operation chain and supply chain. The transition to the CBM might cause severe consequences to social well-being, especially in Vietnam where the GDP is relatively moderate and high development speed due to industrialisation. Hence, the maximisation of stakeholder-centric involvement is crucial to ensure continuous operation. The CBM aligns with the SDGs promoting many advantages for businesses by reaching environmental aspect. However, if the social aspect with humanity elements is neglected, Vietnam would be lagged behind in the development of social welfare. This aspect was justified by Schröder (2020) that the shift to a circular economy will have a direct effect on how social benefits are allocated among nations, businesses, labor unions, and consumers. From 2024, Vietnam's government enforces the Extended Producer Responsibility (EPR) within the business sector requiring companies to extend their responsible activities to tackle the disposed products or to recycle wastes. EPR increases operational costs, however, it opens opportunities for businesses to create more employment and decent working conditions. Schröder (2020) claimed that financial investment could emerge from the public–private partnerships for the collective understanding of stakeholders and the public sector.

7.5 Future Directions

Tran (2023) suggested that CBMs primarily focus on creating, capturing, and delivering value while ensuring that products remain within material closed loops. However, CBMs do not necessarily integrate social and ecological value creation as seen in conventional business models (Tran, 2023). The applicability of CBMs depends on the nature of the business and industry. An inclusive circular economy model, in particular, generates a synergistic effect by integrating circular and inclusive dimensions, thus enhancing business contributions to sustainable development at local, national, and global levels. Moving forward, empirical research is necessary to validate this proposed framework and develop practical implementation guidelines for businesses in underdeveloped and developing countries. Such research would provide actionable insights to facilitate the adoption of inclusive CBMs, enabling developing nations to leapfrog traditional economic models and embrace sustainability as a driver of economic and social progress. By addressing inclusivity and circularity in tandem, businesses can achieve a balanced sustainability approach, ensuring that their impact extends beyond profitability to broader societal and environmental objectives.

7.6 Conclusion

This study underscores the necessity of integrating inclusiveness and circularity into sustainable business models to address pressing societal and environmental challenges. Scholars such as Schoneveld (2020) and Bocken et al. (2014) have emphasized the interplay between these elements and sustainability, reinforcing the need to consider both social values and environmental objectives. While the transition toward a circular economy has gained global attention, it is critical to include human-centric aspects, such as decent work conditions and poverty alleviation, to avoid pitfalls like social washing (Murray et al., 2015; Rizzi et al., 2020; Steenmans & Lesniewska, 2023).

In Vietnam's context, the government has made significant strides in encouraging businesses to adopt a circular economy model. SMEs are beginning to integrate inclusiveness into circular practices, demonstrating a move beyond conventional inclusive toward more comprehensive sustainable strategies (Tran, 2023). Despite these efforts, the number of companies embracing this business model remains limited, highlighting the need for broader implementation and stronger policy frameworks. Without collective action from policymakers, businesses, and stakeholders, social inequalities could persist, undermining the potential benefits of the circular economy.

This research contributes to the conceptualisation of a sustainable business model that synergises inclusivity and circularity, addressing the dual goals of societal and environmental value creation. By embedding inclusiveness within the circular framework, businesses can better align with stakeholder expectations and ensure the long-term viability of their operations. This study also underscores the rising demand for businesses to resolve social issues alongside environmental challenges, suggesting a shift in global expectations for corporate responsibilities.

Highlights

- Sustainable business models will increasingly embed Circular Economy principles and social inclusiveness.
- Closed-loop systems will become central to product and process designs.
- Product lifecycles will be strategically extended to reduce waste and resource use.
- Future business models will prioritize social inclusion, explicitly engaging Bottom of the Pyramid (BoP) communities.
- Affordable products and services will expand market access and address inequality.
- Stakeholder collaboration will intensify, focusing on shared value creation across the ecosystem.
- Local capacity building will grow significantly, enhancing resilience and inclusivity in value chains.

References

Amran, A., & Ooi, S. K. (2014). Sustainability reporting: Meeting stakeholder demands. *Strategic Direction, 30*(7), 38–41. https://doi.org/10.1108/SD-03-2014-0035

Antikainen, M., & Valkokari, K. (2016). A framework for sustainable circular business model innovation. *Technology Innovation Management Review, 6*(7), 5–12. https://doi.org/10.22215/timreview1000

Beckwith, N. (2018). Inclusive business financing—Where commercial opportunity and sustainability coverage. Mandaluyong: Asian Development Bank. https://doi.org/10.22617/TCS189391-2

Bocken, N., Short, S.W., & Evans, P. S. (2014). A literature and practice review to develop sustainable business model archetypes. *Journal of Cleaner Production, 65*, 42–56. https://doi.org/10.1016/j.jclepro.2013.11.039

Bonnell, V., & Veglio, F. (2011). *Inclusive business for sustainable livelihoods.* https://journals.openedition.org/factsreports/840

Boons, F., & Lüdeke-Freund, F. (2012). Business models for sustainable innovation: State-of-the-art and steps towards a research agenda. *Journal of Cleaner Production, 45*, 9–19. https://doi.org/10.1016/j.jclepro.2012.07.007

Butt, A. S., Ali, I., & Govindan, K. (2023). The role of reverse logistics in a circular economy for achieving sustainable development goals: A multiple case study of retail firms. Production Planning & Control. https://doi.org/10.1080/09537287.2023.2197851

Chen, C.-W. (2020). Improving circular economy business models: Opportunities for business and innovation: A new framework for businesses to create a truly circular economy. *Johnson Matthey Technology Review, 64*(1), 48–58. https://doi.org/10.1595/205651320X15710564137538

Corvellec, H., Stowell, A. F., & Johansson, N. (2021). Critiques of the circular economy. *Journal of Industrial Ecology, 26*(2), 421–432. https://doi.org/10.1111/jiec.13187

European Commission. (2008). Waste framework directive. Hämtat från https://environment.ec.europa.eu/topics/waste-and-recycling/waste-framework-directive_en. den 23 January 2023

Ferraz, D., & Pyka, A. (2023). Circular economy, bioeconomy, and sustainable development goals: A systematic literature review. *Environmental Science and Pollution Research.* https://doi.org/10.1007/s11356-023-29632-0

G20 development working group. (2015). G20 inclusive business framework. G20. Hämtat från https://www.inclusivebusiness.net/sites/default/files/inline-files/G20%2BInclusive%2BBusiness%2BFramework_Final.pdf

Geissdoerfer, M., Bocken, N., & JanHultink, E. (2016). Design thinking to enhance the sustainable business modelling process—A workshop based on a value mapping process. *Journal of Cleaner Production, 135*, 1218–1232. https://doi.org/10.1016/j.jclepro.2016.07.020

Geisendorf, S., & Pietrulla, F. (2017). The circular economy and circular economic concepts—a literature analysis and redefinition. *Thunderbird International Business Review, 60*(5), 771–782. https://doi.org/10.1002/tie.21924

Geissdoerfer, M., Bocken, N. M., & Hultink, E. J. (2017). The circular economy—A new sustainability paradigm? *Journal of Cleaner Production, 143*, 757–768. https://doi.org/10.1016/j.jclepro.2016.12.048

Geissdoerfer, M., Vladimirova, D., & Evans, S. (2018). Sustainable business model innovation: A review. *Journal of Cleaner Production, 198*, 401–416. https://doi.org/10.1016/j.jclepro.2018.06.240

Geissdoerfer, M., Pieroni, M. P., Pigosso, D. C., & Soufani, K. (2020). Circular business models: A review. *Journal of Cleaner Production, 277*, 123741. https://doi.org/10.1016/j.jclepro.2020.123741

Ghisellinia, P., Cialani, C., & Ulgiati, S. (2016). A review on circular economy: The expected transition to a balanced interplay of environmental and economic systems. *Journal of Cleaner Production, 116*, 11–32. https://doi.org/10.1016/j.jclepro.2015.09.007

Golja, T., & Požega, S. (2012). Inclusive business-what it is all about? Managing inclusive companies. *International Review of Management and Marketing, 2*(1), 22–42. https://dergipark.org.tr/en/download/article-file/366606

Hoffman, M., Schenck, C. J., & Herbst, F. (2022). Exploring the intersection where business models, a circular economy and sustainability meet in the waste economy: A scoping review. *Sustainability, 14*(6), 3687. https://doi.org/10.3390/su14063687

Hofmann, F. (2019). Circular business models: Business approach as driver or obstructer of sustainability transitions? *Journal of Cleaner Production, 224*, 361–374. https://doi.org/10.1016/j.jclepro.2019.03.115

International Finance Corporation. (2018). Inclusive business at the base of the pyramid. Hämtat från International Finance Corporation: https://www.ifc.org/en/what-we-do/sector-expertise/gender/inclusive-business. den 27 September 2023

Jaakkola, E. (2020). Designing conceptual articles: Four approaches. *AMS Review, 10*(1–2), 18–26. https://doi.org/10.1007/s13162-020-00161-0

Jensen, J. P., Prendeville, S. M., Bocken, N. M., & Peck, D. (2019). Creating sustainable value through remanufacturing: Three industry cases. Journal of Cleaner Production, 218, 304–314. https://https://doi.org/10.1016/j.jclepro.2019.01.301

Kelly, S., Vergara, N., & Bammann, H. (2015). Inclusive business models—The conceptual framework. In S. Kelly, N. Vergara, & H. Bammann (Eds.), Nclusive business models—Guidelines for improving linkages between producer groups and buyers of agricultural produce. FAO. Hämtat från https://www.fao.org/publications/card/en/c/1b8e4cb1-d29f-45f7-b331-3655507cf466/

Kolk, A., Rivera-Santos, M., & Rufín, C. (2013). Reviewing a decade of research on the "Base/Bottom of the Pyramid" (BOP) concept. *Business & Society, 53*(3), 338–377. https://doi.org/10.1177/0007650312474928

Korhonen, J., Honkasalo, A., & Seppälä, J. (2018). Circular economy: The concept and its limitations. *Ecological Economics, 143*, 37–46. https://doi.org/10.1016/j.ecolecon.2017.06.041

Klasen, S. (2010). Measuring and monitoring inclusive growth: Multiple definitions, open questions, and some constructive proposals. Asian Development Bank Sustainable Development Working Paper Series. https://www.adb.org/sites/default/files/publication/28492/adb-wp12-measuring-inclusive-growth.pdf

Kurnia, S., Alamsyahbana, M. I., Chartady, R., Arifin, S. V., & Sesaria, M. I. (2023). Circular solutions for decent work and economic growth: Lessons from sustainable development goals (SDG) 8. *Academia Open, 8*(1), 1–13. https://doi.org/10.21070/acopen.8.2023.6657

Lewandowski, M. (2016). Designing the business models for circular economy—Towards the conceptual framework. *Sustainability, 8*(1), 43. https://doi.org/10.3390/su8010043

Likoko, E., & Kini, J. (2017). Inclusive business—A business approach to development. *Current Opinion in Environmental Sustainability, 24*, 84–88. https://doi.org/10.1016/j.cosust.2017.03.001

Linder, M., & Williander, M. (2015). Circular business model innovation: Inherent uncertainties. *Business Strategy and the Environment, 26*(2), 182–196. https://doi.org/10.1002/bse.1906

Mangnus, E. (2019). How inclusive businesses can contribute to local food security. *Current Opinion in Environmental Sustainability, 41*, 69–73. https://doi.org/10.1016/j.cosust.2019.10.009

Mishra, J. L., Hopkinson, P. G., & Tidridge, G. (2018). Value creation from circular economy-led closed loop supply chains: A case study of fast-moving consumer goods. *Production Planning & Control, 29*(6), 509–521. https://doi.org/10.1080/09537287.2018.1449245

Murray, A., Skene, K., & Haynes, K. (2015). The circular economy: An interdisciplinary exploration of the concept and application in a global context. *Journal of Business Ethics, 140*, 369–380. https://doi.org/10.1007/s10551-015-2693-2

Nandi, S., Hervani, A. A., & Helms, M. M. (2020). Circular economy business models—Supply chain perspectives. *IEEE Engineering Management Review, 48*(2), 193–201. https://doi.org/10.1109/emr.2020.2991388

Nguyen, H. Q., Dang, T. H., Nguyen, J., Ha, J., Phan, D. T., & Pham, P. T. (2021). Một số vấn đề về phát triển kinh tế tuần hoàn tại Việt Nam [Some issues on the development of circular economy in Vietnam]. *Tạp chí Khoa học và Công nghệ Việt Nam, 2021*(1+2A), 49–51. https://scholar.dlu.edu.vn/thuvienso/bitstream/DLU123456789/257898/1/BaiTapChi_JDZ8DTQY.pdf

Nguyen, H. N, Nguyen, T. C., & Tran, V. Y. (2020). Mối quan hệ giữa Tăng trưởng xanh, Kinh tế xanh, Kinh tế tuần hoàn và Phát triển bền vững [The Relationship Between Green Growth, Green Economy, Circular Economy and Sustainable Development]. *Tạp chí Nghiên cứu kinh tế, Số 5*(504)-Tháng 5/2020, 47–56. https://www.researchgate.net/profile/Nam-Hoang-Nguyen/publication/326032706_Phan_biet_khai_niem_kinh_te_nau_kinh_te_xanh_tang_truong_xanh_kinh_te_tuyen_tinh_kinh_te_tuan_hoan_va_phat_trien_ben_vung_Terminological_distinctions_between_khai-niem-kinh-te-nau-kinh-te-xanh-tang-truong-xanh-kinh-te-tuyen-tinh-kinh-te-tuan-hoan-va-phat-trien-ben-vung-Terminological-distinctions-between-Brown-economy-Green-economy-Green-gr.pdf

Nikolaou, I. E., Jones, N., & Stefanakis, A. (2021). Circular economy and sustainability: The past, the present and the future directions. *Circular Economy and Sustainability, 1*, 1–20. https://doi.org/10.1007/s43615-021-00030-3

Nußholz, J. L. K. (2018). A circular business model mapping tool for creating value from prolonged product lifetime and closed material loops. *Journal of Cleaner Production, 197*(Part 1), 185–194. https://doi.org/10.1016/j.jclepro.2018.06.112

Oberlack, C. T. B., Zambrino, L., Bruelisauer, S., Solar, J., Villar, G., Ramírez, M. (2023). With and beyond sustainability certification: Exploring inclusive business and solidarity economy strategies in Peru and Switzerland. *World Development, 165*, 106187. https://doi.org/10.1016/j.worlddev.2023.106187

Phan, C. T., Vo, T. T., & Vo, D. T. (2023). Can microcredit reduce vulnerability to poverty? Evidence from rural Vietnam. *Review of Development Economics, 27*(1), 608–629. https://doi.org/10.1111/rode.12951

Pieroni, M. P., McAloone, T. C., & Pigosso, D. C. (2019). Business model innovation for circular economy and sustainability: A review of approaches. *Journal of Cleaner Production, 215*, 198–216. https://doi.org/10.1016/j.jclepro.2019.01.036

Pla-Julián, I., & Guevara, S. (2019). Is circular economy the key to transitioning towards sustainable development? Challenges from the perspective of care ethics. *Futures, 105*, 67–77. https://doi.org/10.1016/j.futures.2018.09.001

Pollard, J., Osmani, M., Grubnic, S., Díaz, A. I., Grobe, K., Kaba, A., Panchal, R. et al. (2023). Implementing a circular economy business model canvas in the electrical and electronic manufacturing sector: A case study approach. *Sustainable Production and Consumption, 36*, 17–31. https://doi.org/10.1016/j.spc.2022.12.009

Potting, J., Hekkert, M. P., Worrell, E., & Hanemaaijer, A. (2017). Circular economy: *Measuring innovation in the product chain*. https://researchportal.uu.nl/files/40538049/Circular.pdf

Ranieri, R., & Almeida Ramos, R. (2013). Inclusive growth: Building up a concept, Working papers 104, International Policy Centre for Inclusive Growth. https://www.econstor.eu/bitstream/10419/71821/1/736729240.pdf

Rezaee, Z. (2016). Business sustainability research: A theoretical and integrated perspective. *Journal of Accounting Literature, 36*(1), 48–64. https://doi.org/10.1016/j.acclit.2016.05.003

Rezaee, Z., Zhen, K., & Ha, M. (2017). Progress toward business sustainability in asia in the aftermath of 2015 Hong Kong stock exchange requirements. *International Journal of Sustainability Management and Information Technologies, 3*(4), 40–45. https://doi.org/10.11648/j.ijsmit.20170304.12

Rizzi, F., Gusmerotti, N., & Frey, M. (2020). How to meet reuse and preparation for reuse targets? Shape advertising strategies but be aware of "social washing". *Waste Management, 101*, 291–300. https://doi.org/10.1016/j.wasman.2019.10.024

Romero, C. A. T., Castro, D. F., Ortiz, J. H., Khalaf, O. I., & Vargas, M. A. (2021). Synergy between circular economy and industry 4.0: A literature review. *Sustainability, 13*(8), 4331. https://doi.org/10.3390/su13084331

Rossi, E., Bertassini, A. C., Ferreira, C. d., Amaral, W. A., & Ometto, A. R. (2020). Circular economy indicators for organizations considering sustainability and business models: Plastic, textile and electro-electronic cases. *Journal of Cleaner Production, 247*, 119–137. https://doi.org/10.1016/j.jclepro.2019.119137

Ros-Tonen, M. A., Bitzer, V., Laven, A., Leth, D. O., Leynseele, Y. V., & Vos, A. (2019). Conceptualizing inclusiveness of smallholder value chain integration. *Current Opinion in Environmental Sustainability, 41*, 10–17. https://doi.org/10.1016/j.cosust.2019.08.006

Seelos, C., & Mair, J. (2004). Social entrepreneurship: Creating new business models to serve the poor. *Business Horizons, 48*(3), 241–246. https://doi.org/10.1016/j.bushor.2004.11.006

Schoneveld, G. C. (2020). Sustainable business models for inclusive growth: Towards a conceptual foundation of inclusive business. *Journal of Cleaner Production, 277*. https://doi.org/10.1016/j.jclepro.2020.124062

Schröder, P. (2020). *Promoting a just transition to an inclusive circular economy*. Chatham House. https://doi.org/10.13140/RG.2.2.24974.59204

Schulte, O., Mumber, J., & Nguyen, T. T. (2023). Agricultural commercialisation, asset growth and poverty in rural Vietnam. *Agricultural and resource economics, 67*(3), 388–416. https://doi.org/10.1111/1467-8489.12517

Sehnem, S., Pandolfi, A., & Gomes, C. (2019). Is sustainability a driver of the circular economy? *Social Responsibility Journal, 16*(3), 329–347. https://doi.org/10.1108/SRJ-06-2018-0146

SNV; World Business Council for Sustainable Development. (2008). Inclusive business: Profitable business for successful development. The Hague: SNV and World Business Council for Sustainable Development. Hämtat från https://docs.wbcsd.org/2008/03/Inclusive_Business.pdf

SNV; World Business Council for Sustainable Development. (2011). Iclusive business: Creating value in Latin America. The Hague: SNV and World Business Council for Sustainable Development. Hämtat från https://www.wbcsd.org/contentwbc/download/2898/36354/1

Speelman, E. N., Rodela, R., Doddema, M., & Ligtenberg, A. (2019). Serious gaming as a tool to facilitate inclusive business; A review of untapped potential. *Current Opinion in Environmental Sustainability, 41*, 31–37. https://doi.org/10.1016/j.cosust.2019.09.010

Steenmans, K., & Lesniewska, F. (2023). Limitations of the circular economy concept in law and policy. *Frontiers in Sustainability, 4*. https://doi.org/10.3389/frsus.2023.1154059

Suriyankietkaew, S., & Avery, G. (2016). Sustainable leadership practices driving financial performance: Empirical evidence from Thai SMEs. *Sustainability, 8*(4), 327. https://doi.org/10.3390/su8040327

Svensson, G., Ferro, C., Hogevold, N., Padin, C., & Varela, J. C. (2018). Developing a theory of focal company business sustainability efforts in connection with supply chain stakeholders. *Supply Chain Management, 23*(1), 16–32. https://doi.org/10.1108/SCM-12-2015-0461

Teece, D. (2010). Business models, business strategy and innovation. *Long Range Planning, 43*(2–3), 172–194. https://doi.org/10.1016/j.lrp.2009.07.003

The Inclusive Business Action Network. (den 22 September 2021). INCLUSIVE BUSINESS FEATURES. Hämtat från The Inclusive Business Action Network (iBAN): https://www.inclusivebusiness.net/node/5432. den 15 January 2024

Todeschini, B. V., Cortimiglia, M. N., Callegaro-de-Menezes, D., & Ghezzi, A. (2017). Innovative and sustainable business models in the fashion industry: Entrepreneurial drivers, opportunities, and challenges. *Business Horizons, 60*(6), 759–770. https://doi.org/10.1016/j.bushor.2017.07.003

Tran, T. P., Nguyen, N. Q., Nguyen, C. D., Huynh, V. C., Hoang, D. H., & HungHoang, T. (2023). Livelihood vulnerability to climate change: Indexes and insights from two ethnic minority communities in Central Vietnam. *Environmental Challenges, 10*, 100666. https://doi.org/10.1016/j.envc.2022.100666

Tran, A. T. (2023). Exploring the circular economy embedded in the business model. The perspective of Vietnam. *Studia Universitatis Babes-Bolyai Negotia*, 67–92. https://doi.org/10.24193/subbnegotia.2023.3.03

UNDP. (2010). *Brokering inclusive business models*. UNDP. New York. https://www.local2030.org/library/view/63

United Nations Development Programme. (2013). *Business solutions to poverty: How inclusive business models create opportunities for all in emerging Europe and Central Asia.* UNDP. Hämtat från https://info.undp.org/docs/pdc/Documents/SVK/00057199_GIM%20Report%20Laid%20out2.pdf

United Nations Development Programme. (2016). *Brokering inclusive business models.* Brokering Inclusive Business Models.

Velte, C. J., & Steinhilper, R. (2016, June). Complexity in a circular economy: A need for rethinking complexity management strategies. In *Proceedings of the World Congress on Engineering* (Vol. 29). London, UK.

Vermunt, D., S.O. Negro, P. V., Kuppens, D., & Hekkert, M. (2019). Exploring barriers to implementing different circular business models. *Journal of Cleaner Production, 222*, 891–902. https://doi.org/10.1016/j.jclepro.2019.03.052

Wells, P. (2015). Economies of scale versus small is beautiful: A business model approach based on architecture, principles and components in the beer industry. *Organization & Environment, 29*(1), 36–52. https://doi.org/10.1177/1086026615590882

Wells, P., & Seitz, M. (2005). Business models and closed-loop supply chains: A typology. *Supply Chain Management, 10*(4), 249–251. https://doi.org/10.1108/13598540510612712

Whalen, K. A. (2019). Three circular business models that extend product value and their contribution to resource efficiency. *Journal of Cleaner Production, 226*, 1128–1137. https://doi.org/10.1016/j.jclepro.2019.03.128

Witjes, S., & Lozano, R. (2016). Towards a more circular economy: Proposing a framework linking sustainable public procurement and sustainable business models. *Resources, Conservation and Recycling, 112*, 37–44. https://doi.org/10.1016/j.resconrec.2016.04.015

Anh Tuan Tran is currently working at the Department of Development and Cooperation at the University of Economics and Law as a coordinator for international partnerships and projects. Furthermore, he is a junior researcher in International Business and Political Economy. He specialises in Entrepreneurship and Business Studies. His background is rooted in the study of international economic relations and business. His research interests include business and national policy analysis, as well as International Economic Relations, among others. Recently, he has broadened his research scope to encompass an interdisciplinary approach. To that end, he is conducting research in the field of international business, focusing on Private–Public Partnerships, corporate political activities, and corporate social responsibility. He has published several relevant articles, such as those concerning the soft power of firms and business diplomacy. In the area of business research, he is also collaborating with colleagues on the circular economy and sustainable business models for firms, as well as just transition initiatives in Vietnam.

Thu Ha Nguyen is the principal lecturer with 27 years of experience in teaching at the Industrial University of Ho Chi Minh City and Van Lang University. Her research interests include cultural studies, cultural and public diplomacy, linguistics, teaching methods, and educational innovation. Recently, she has extended her research interest to business development in the cultural industry.

Anh Tuan Nguyen serves as a lecturer in the Department of Hospitality Management at the School of Hospitality and Tourism, Hue University of Vietnam. He obtained his master's degree from Budapest Business University, where his thesis focused on community-based tourism development. His research interests include sustainable business management and reducing the carbon footprint of businesses. He is also leading several publications on sustainability for heritage city development and sustainable tourism.

Part III
Technological Trends and Market Dynamics

Chapter 8
A Systematic Literature Review of Extended Reality in Marketing: An Asian Perspective with a Focus on India

Clusters of Future Studies: Past and Futures; Technological Trends

Nishant Pandey, Ankit Kumar Sinha, and Anand Kumar

Abstract Extended reality (XR) is gaining popularity among researchers, and they are investigating its impact on marketing. This comprehensive study delves into the evolving realm of extended reality-based technologies in marketing, with a targeted examination of the Asia Pacific region, with India as a case. The study did a systematic literature review that included 17 peer-reviewed articles published from 2005 to 2023 in the Scopus database based on the keyword search ("Extended reality" OR "XR" OR "Mixed reality" OR "Metaverse") AND ("Marketing" OR "Promotion" OR "Branding" OR "Sales" OR "Customer Relationship" OR "Consumer Behaviour." The selected articles corpus helped this study gain insight into discerning theories, barriers, advantages, and disadvantages associated with integrating extended reality in marketing. The review focused on the in-depth case study of pioneering entities, IKEA and Lenskart, situated in India, providing illuminating perspectives on effective implementation strategies. The findings provide valuable guidance for businesses contemplating the incorporation of extended reality in their marketing strategies. Notably, the study concludes by proposing future research directions within this dynamic domain, fostering a roadmap for further scholarly inquiry. This study contributes to the understanding of the multifaceted landscape of extended reality in marketing and provides the outlook for informed decision-making and strategic planning in the business for future researchers and managers.

N. Pandey (✉) · A. K. Sinha
Indian Institute of Management Ranchi, Ranchi, Jharkhand, India
e-mail: nishant.pandeyi22@iimranchi.ac.in

A. K. Sinha
e-mail: ankit.kumarsinha22ph@iimranchi.ac.in

A. Kumar
Institute of Management, Nirma University, Ahmedabad, Gujarat, India
e-mail: anand.kumar@nirmauni.ac.in

M. S. Balaji et al. (eds.), *Future of Business in Asia*,
https://doi.org/10.1007/978-3-031-95940-0_8

Keywords Extended reality · Lenskart · IKEA marketing · Asia Pacific · India · Systematic literature review

8.1 Introduction

Modern marketers predict that extended reality-based technology will become the new arena for marketing (Alcañiz et al., 2019). Extended reality (XR) serves as a comprehensive umbrella term encompassing a spectrum of immersive technologies, comprising augmented reality (AR), virtual reality (VR), and mixed reality (MR) (Thomas Alsop, 2023). It is transforming the user's digital world interaction experience. The AR and VR market in the United States is indeed set for substantial growth. A forecasted revenue increase to $5.7 billion by 2027, reflecting a +66.51% growth rate, points to accelerating adoption across industries. It is projected to reach a new peak of 14.24 billion US dollars in 2027 after the tenth consecutive year of growth, it is noteworthy that the revenue in the AR & VR market has exhibited a consistent upward trend over the past years (Thomas Alsop, 2024). The inclination of marketers due to the expected growth has led us to investigate the domain. Through this domain investigation, authors can identify the knowledge gap in the literature. The current literature is scattered and needs the knowledge integration of the current body of knowledge in extended reality-based technologies and their usage in marketing to give a holistic overview of this technology's existing state and future research agenda.

A systematic literature review is proposed to address this knowledge gap. It aims to furnish an in-depth overview of the present research landscape, elucidating the existing challenges and delineating potential directions for future research endeavours (Martínez-Rojas et al., 2018). The literature review started with the keyword search on the Scopus database and was followed by the inclusion and exclusion criteria prescribed in the PRISMA (Preferred Reporting Items for Systematic Reviews and Meta-Analyses) framework proposed by (Paul et al., 2021). Through the initial search on the Scopus database, 17 articles have been identified and finalised for the study. The study identifies the significant theories utilised by previous researchers in understanding XR-based technologies. Further analysis in the Asia Pacific region with a special focus in India is undertaken so as to provide context and linkage from theory to practice in one of the two one-billion population country in the region. The extant literatures were analysed leading to the identification of various advantages and disadvantages of adopting XR-based technologies in marketing for organisations and asian countries.

The review was instrumental in identifying two of the most prominent companies in India that have been at the forefront of extended reality-based technologies in their marketing and operations i.e., IKEA and Lenskart. In light of their pioneering efforts in adopting XR marketing, a detailed case study on these two organisations is performed to gain a deeper understanding of their role in this domain. The findings from the case study provided valuable insights into the effective implementation of extended reality-based technologies in marketing and operations, which could serve as a useful reference for other businesses looking to adopt similar strategies. Additionally, the study identified future research directions that could help researchers further explore the domain of XR in the marketing domain.

The book chapter has been further divided into sections for better comprehension. Section 8.2 provides the research methodology adopted in this study. Section 8.3 deals with the review findings, which include XR technologies adoption, barriers, advantages, and disadvantages of XR marketing, and the identification of various theories in the extended reality-based marketing literature. Section 8.4 consists of case studies on the applications of XR technologies in India focusing on IKEA and Lenskart. Section 8.5 provides significant future direction to the researchers in this domain. Lastly, Sect. 8.6 concludes the review study.

8.2 Review Methodology

The study started with a keyword search on the Scopus database, which possesses an extensive repository of rigorously peer-reviewed scholarly articles (Chadegani et al., 2013). Keywords were utilised to have exclusive peer-reviewed articles, including book chapters that focused on our objective, which is the integration of literature in XR and marketing. Therefore, we have used the AND operator in our keyword search query, shown in Table 8.1.

The study utilised the PRISMA framework for inclusion and exclusion, as shown in Fig. 8.1. Through an extensive search of the Scopus database, authors have identified 387 documents. Additionally, authors have manually searched journal websites of some of the prominent journals publishing in the extended reality domain. The journals included in the final are the *International Journal of Information Management, Journal of Management Information Systems, MIS Quarterly, Electronic Markets, Electronic Commerce Research, International Journal of Human–Computer Interaction, and Journal of Decision Systems.* The journal search identifies 259 documents. The duplicate documents are excluded to reach the number of 587 unique documents.

Table 8.1 Scopus search query of XR (extended reality) in marketing

Search query	
Search date	20 December 2023
Search period	1991 to September 2023
Search topic	("Extended reality" OR "XR" OR "Mixed reality" OR "Metaverse") AND ("Marketing" OR "Promotion" OR "Branding" OR "Sales" OR "Customer Relationship" OR "Consumer Behaviour")
Document type	Only include peer-reviewed journal and book chapters
Document language	English

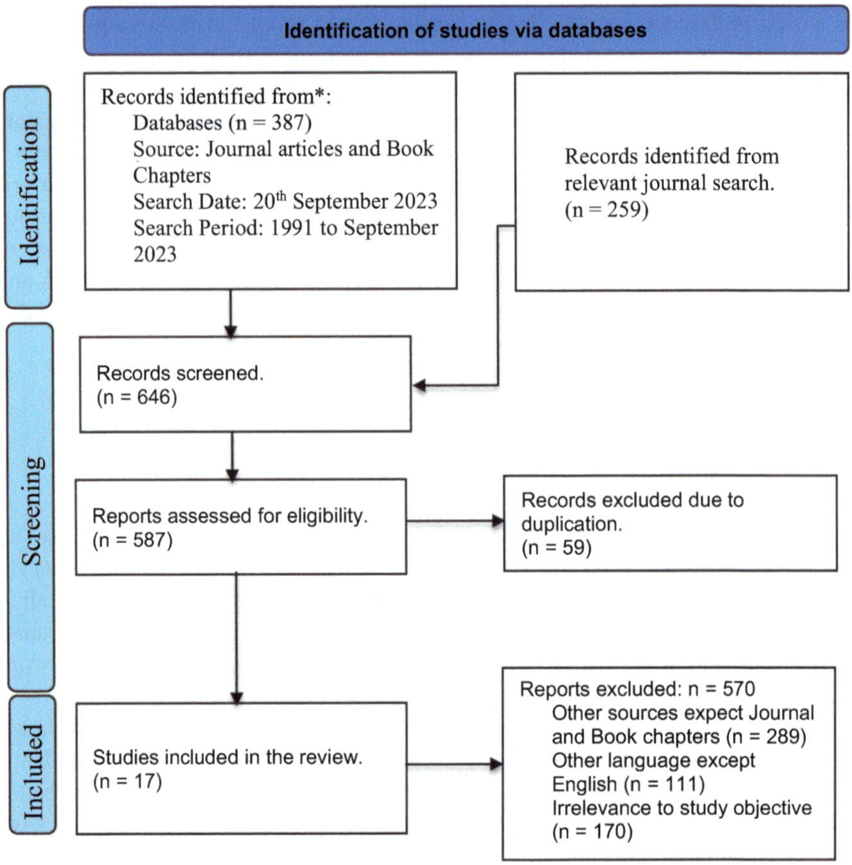

Fig. 8.1 PRISMA flow diagram of review on XR (extended reality) in marketing

The exclusion criteria of document type are applied, and only Journal articles and book chapters are selected. Due to the comprehension of the authors and readers of this study, authors have mutually decided only to include the documents in the English language. After the previously imposed inclusion and exclusion criteria, authors were left with a total of 187 documents, which were manually investigated by the authors through title, abstract and keyword. Through the intercoder reliability was checked and the score revealed that the chosen articles aligned with the aim of the study, finally 17 articles were chosen for the final study. A group of senior experienced scholars have been consulted to verify the effectiveness of chosen documents in meeting the study objectives. They reported that the included study matches our study objectives.

8.3 Findings

This study conducted a systematic literature review to examine extended reality (XR) in marketing, branding, sales, customer relationship, and consumer behaviour, focusing on articles published between 2005 and 2023 in the Scopus database. Using keywords such as "Extended reality," "XR," "Mixed reality," "Metaverse," combined with "Marketing," "Promotion," "Branding," "Sales," "Customer Relationship," and "Consumer Behaviour," the search identified 17 peer-reviewed articles. Studies with an Indian context were prioritised to align with the objective to understand XR's impact in India, ensuring relevance to the region. The list of the identified studies, along with the key findings, are mentioned in Table 8.2.

8.3.1 Adoption of XR Technologies

The adoption of XR is driven by its potential to revolutionise customer experiences, enhance employee training, and create innovative marketing strategies. Businesses are leveraging XR to blur the boundaries between physical and digital environments, offering personalised and engaging experiences (Arya et al., 2024; Yadav et al., 2024). In marketing, XR enables brands to craft immersive storytelling, leading to higher customer engagement and loyalty. XR is expected to see broader integration, particularly in sectors like retail, healthcare, and education, where it holds the promise of transforming user interaction and operational efficiency.

8.3.2 Barriers to Adoption of XR Technologies

Contemporary challenges within the domain of XR are meticulously delineated (Flotyński, 2022). The constraint emanating from the restricted availability of aspect-oriented libraries compatible with game engines poses a notable hindrance to the seamless development of immersive XR experiences. Moreover, the perpetual evolution of XR technologies necessitates continuous optimisation and extension of implementations (Stary, 2023). The imperative for domain experts to comprehend and analyse behaviour within XR environments underscores the indispensability of specialised skills in deciphering user interactions within these immersive spaces.

A pivotal barrier surfaces in the form of a dearth of XR skills within organisations, thereby necessitating external consulting or training for the effective implementation of XR solutions. The palpable influence of competitor pressure on immersive technology integration compels companies to contemplate XR adoption, particularly within customer-facing processes. Furthermore, cultural intricacies, encapsulating stakeholder preferences and readiness for XR technologies, emerge as substantial

Table 8.2 List of articles in the final corpus of XR (extended reality) in marketing (along with key findings)

Reference	Research context	Theoretical perspective	Key findings
Stary (2023)	Exploration of integrating Metaverse and XR technologies into everyday applications to address sustainability by optimising resource consumption before implementation	Utilises digital process twin technologies to model behaviour-centred intelligence of socio-technical cyber-physical systems within Metaverse/XR environments	Demonstrates that accurately capturing user roles and cyber-physical components through behaviour models allows for precise assessment of environmental, social, and economic impacts of different design alternatives in Metaverse/XR applications
Shaikh et al. (2022)	Investigates the maturation of XR solutions in healthcare, exploring novel use cases and therapeutic applications	Technological innovation and medical service delivery	Market growth: XR market anticipated to reach $209 billion by 2022. Promising therapeutic applications and improved training for various medical scenarios
Rauschnabel et al. (2022)	Exploration of AR in marketing, aiming to establish AR marketing as a holistic and strategic subdiscipline	Strategic marketing theory, customer journey models, framework development	Defined "augmented reality marketing" as a novel marketing subdiscipline. Proposed the BICK FOUR framework and provided insights from managers on AR marketing practices

(continued)

Table 8.2 (continued)

Reference	Research context	Theoretical perspective	Key findings
Ortega-Gras et al. (2023)	Exploration of XR technologies (VR, AR, MR) as key enablers of Industry 4.0 and the metaverse, focusing on vocational education and training (VET) in the W&F sector	Industry 4.0 & digital transformation with a focus on XR as a transformative tool in modern industrial settings	XR tools evaluation, successful applications, and stakeholder insights with benefits such as cost savings and support for the circular economy
Lee (2012)	Exploration of various educational and training methods, focusing on AR integration	Technology adoption in educational settings	AR significantly alters the timing and location of education, enhancing efficiency
Kumar and Srivastava (2022)	Examines the role of AR in online impulse behaviour for high-involvement products	Extends literature on AR and online impulse behaviour from a psychological perspective	AR virtual try-on significantly influences online impulse behaviour, with flow and spatial presence as mediating factors
Koohang et al. (2023)	Examines the adoption and impact of the metaverse across various industries	Not specified	Overview of metaverse technologies, challenges, and research agenda across industries
Kohli et al. (2022)	Examines the convergence of brain-computer interfaces (BCIs) and XR in smart cities	Human–machine interaction for applications like rehabilitation, navigation, and smart city control	BCI and XR combination shows promise in enhancing smart city applications across rehabilitation, robotics, and home control

(continued)

Table 8.2 (continued)

Reference	Research context	Theoretical perspective	Key findings
Jim et al. (2023)	Overview of Metaverse applications, challenges, and solutions, focusing on establishing trust in the Metaverse ecosystem	Examines Metaverse applications, challenges, trustworthiness, and solutions through existing literature	Identified application areas, challenges, and solutions with a future research agenda for trustworthiness in the Metaverse
Jalo et al. (2022)	Investigates awareness, limitations, and adoption of AR and VR in European industrial companies, with focus on SMEs versus large companies	Technology–organisation–environment (TOE) framework	Larger companies adopt XR more than SMEs. No significant difference between AR and VR perceptions
Gong et al. (2021)	Investigating VR applications in enhancing creativity during brainstorming	Creativity enhancement via technology	Categories (avatars, environments, interfaces, applications); trends and measures for evaluating creativity
Flotyński (2022)	Exploring user and object interactions within XR for domain-focused analysis without requiring technical 3D expertise	Semantic representation to enhance XR environments for high-level analysis	Enables efficient, non-technical development of explorable XR environments for domain experts
Burton and Schlieman (2021)	Examines consumer responses to 360° video as a platform for sponsorship and brand marketing	Extended reality (XR) marketing	360° video enhances user experience and brand perception, creating a viable medium for sponsors

determinants influencing the adoption process, particularly within global operations marked by diverse stakeholders (Jalo et al., 2022).

Delving into challenges impinging on XR market maturity, the absence of standardised technical and legal criteria, coupled with a clear workflow for material development, impedes mass-market maturity (Ortega-Gras et al., 2023). Additionally, the limited functionality and accessibility of extant XR products pose formidable obstacles to widespread consumer adoption. These challenges collectively underscore the multifaceted nature of impediments faced in the adoption and maturation of XR technologies across diverse domains. Complementing these challenges, the potential barriers to user acceptance of XR technologies, including unfamiliarity, privacy concerns, or the potential for motion sickness in virtual environments, are illuminated (Ibrahim et al., 2023). This layer of challenges accentuates the imperative of addressing user-centric concerns for the efficacious adoption of XR.

Furthermore, the intricacy of content creation in XR is underscored (Ibrahim et al., 2023). The development of high-quality and engaging XR content necessitates specialised skills and resources, acting as a formidable barrier for organisations with limited expertise in this domain. The imperative of bridging this skill gap and facilitating more straightforward content creation is underscored as pivotal for fostering the development of compelling XR experiences. In this erudite exploration of challenges, the amalgamation of cited research profoundly contributes to a nuanced comprehension of the multifaceted hurdles inherent in the adoption and maturation of XR technologies (Batat & Hammedi, 2023; Flotyński, 2022; Koohang et al., 2023).

8.3.3 Advantages of XR Technologies

XR technologies have emerged as transformative forces in recent years, encompassing augmented, virtual, and mixed reality. These technologies are reshaping industries with immersive capabilities, leading to enriched customer experiences, improved decision-making, and innovative educational approaches. This exploration will delve deeper into XR's far-reaching impact across sectors (Koohang et al., 2023).

The enhanced user experience and interaction within XR environments provide various advantages to organisations; for example, a noted potential for in-depth analysis of user and object behaviour provides valuable insights applicable to training, design, marketing, and other domains (Flotyński, 2022). Increased collaboration opportunities, particularly beneficial for SMEs, have been opened up by XR adoption. The enhancement of customer engagement leading to improved satisfaction has been attributed to XR technologies (Batat & Hammedi, 2023).

The benefits of XR in terms of user engagement and interaction, with a note on the potential for creating unique and stimulating brand experiences, along with the ability to reach a broad audience through accessible media like smartphones and laptops, have been emphasised (Burton & Schlieman, 2021). In education, the improved learning experience facilitated by XR technologies has been highlighted. This enables students to understand abstract concepts better, collaborate effectively with teachers and peers, and foster increased interest and motivation (Ortega-Gras et al., 2023). Positive student recommendations and a preference for XR-based teaching further highlight the transformative impact of these technologies on education.

XR technologies contribute significantly to various sectors, enhancing user experiences, fostering collaboration, and providing innovative solutions for marketing and education. The transformative impact of these advancements on industries is acknowledged, with the potential to improve overall efficiency and effectiveness (Ibrahim et al., 2023).

8.3.4 Disadvantages of XR Technologies

The emergence of extended reality (XR) has sparked significant interest in various industries, including retail marketing. However, data security and privacy have become a concern in this area. Tracking customer emotions and reactions in XR raises ethical considerations, underscoring the need for stringent safeguards to protect user data (Jim et al., 2023). Despite the potential benefits of XR, challenges related to affordability, accessibility, interoperability, and technological limitations pose significant obstacles to widespread adoption. Addressing these challenges is crucial for creating a more inclusive and accessible XR landscape. Moreover, there is a need for further research to comprehend consumer behaviour and preferences within the metaverse environment and to assess the effectiveness of marketing strategies in this emerging space (Tan et al., 2023). This underscores the evolving nature of XR and the need for continuous investigation to inform strategic marketing approaches.

The implementation of XR technologies faces hurdles related to limited technical skills, domain-oriented analysis, and representing temporal properties and behaviour in XR environments (Flotyński, 2022). Advancements in tool development and user-friendly interfaces are necessary to overcome these challenges. The high initial costs associated with XR technologies present a notable barrier to adoption, especially for Small and Medium Enterprises (SMEs) (Batat & Hammedi, 2023). Addressing the financial aspect is crucial for wider XR integration, particularly for smaller businesses looking to leverage immersive technologies.

Employee resistance stands out as a critical limiting factor in XR adoption, necessitating strategies to engage and involve employees in the adoption process (Khanra et al., 2021). Change management strategies are necessary to foster a positive attitude towards XR technologies within organisations. The under-researched areas in sponsorship scholarship highlight gaps in current knowledge (Burton & Schlieman, 2021). Further exploration in these areas can provide valuable insights into the strategic utilisation of XR in sponsorship. Funding and infrastructure challenges remain significant obstacles to the broader implementation of XR technologies, indicating the need for addressing financial and logistical considerations (Ortega-Gras et al., 2023). Overcoming these challenges is crucial for the widespread integration of XR into various sectors. Technological barriers, encompassing the significant investment required in hardware, software, and expertise, pose challenges to the smooth adoption of

XR technologies (Ibrahim et al., 2023). Compatibility issues with existing systems further contribute to the complexities faced in the integration process.

Finally, user acceptance of XR technologies may be hindered by factors such as unfamiliarity, privacy concerns, or potential motion sickness in virtual environments. Additionally, the complexity of content creation in XR, demanding specialised skills and resources, serves as a barrier for organisations with limited expertise in this domain (Miles & Snow, 1984). Bridging this skill gap and facilitating easier content creation is crucial for promoting the development of engaging XR experiences.

8.3.5 Theories Used in XR Literature

The authors thoroughly examined the literature and found that there are several theories that previous researchers have used to understand XR (extended reality). Some of the prominent theories used in the literature have been shown in Table 8.3.

Table 8.3 Theories used in XR (extended reality) literature

Theory	Reference	Description of theory
Co-creation theory	Gursoy et al. (2023); Galvagno and Dalli (2014)	"Co-creation is the joint, collaborative, concurrent, peer-like process of producing new value, both materially and symbolically"
Media richness theory	de Amorim et al. (2022); Staw and Cummings (1984)	"Media is "rich" when it is able to provide multiple information cues, immediate feedback and uses very varied language"
Social identity theory	Brown (2000)	"Individuals seek to maintain a positive social identity by a process of categorisation"
Organisational learning theory	Cummings and Teng (2006)	Organisational learning theory explores how organisations acquire, share, and apply knowledge to adapt and improve over time. It emphasises continuous learning at individual, group, and organisational levels to enhance performance and maintain competitiveness
Flow theory	Brannon et al. (2021)	"When experiencing the notion of flow, individuals often enter into a state where they are completely switched off to the outside (real) world and become so engrossed in an activity that they feel like they are encountering a natural, and highly enjoyable, out of body experience."

8.4 Case Studies on Extended Reality in India

To examine the adoption and usage patterns of XR in India, this study conducted two case studies on IKEA India and Lenskart, both recognised as pioneers in XR innovation within the Indian market.

8.4.1 Case Study of IKEA, India

The adoption of XR technologies by IKEA in the Indian market can be considered as a significant step towards retail transformation. With the aim to cater to the diverse preferences of the Indian consumer base, IKEA marketing is using XR applications to enhance its marketing strategies. By utilising AR, customers can now virtually explore and visualise furniture and home decor offerings, thus, overcoming the traditional challenges related to size and fit uncertainties (Rauschnabel et al., 2022). The introduction of virtual room planning has been a noteworthy feature, which empowers customers to visualise and organise IKEA products in their own homes. Furthermore, IKEA has effectively incorporated XR experiences within its physical stores through interactive kiosks equipped with VR headsets, which provide virtual tours and inspiration. This innovative approach has not only boosted customer confidence in purchasing decisions but also positioned IKEA as a leader in merging technology with traditional retail practices (Zaman et al., 2024). The immersive and personalised shopping experience, thus, delivered by IKEA, transcends conventional boundaries and is a testament to its commitment to customer satisfaction.

8.4.2 Case Study of Lenskart, India

Lenskart is a prominent eyewear brand in the Indian market that has strategically utilised XR technology to revolutionise the eyewear shopping experience (Kumar & Srivastava, 2022). Recognising the inherent challenges associated with online eyewear purchases, Lenskart has proactively integrated XR technologies, specifically focusing on AR, to address concerns related to fit, style, and overall aesthetics. The Lenskart mobile application is at the forefront of this technological integration, empowering customers with the ability to try on a diverse array of eyeglass frames virtually (Kumar and Pathak, 2018). By seamlessly integrating AR technology into the Lenskart app, customers can visualise how different frames complement their facial features in a dynamic and interactive manner. This innovative approach serves as a solution to the longstanding challenges of online eyewear shopping (Imtiaz et al., 2017). The AR-driven virtual try-on feature enhances the online shopping experience, bridging the gap between the digital and physical realms. This not only overcomes barriers to online eyewear shopping but also positions Lenskart as an industry trailblazer, showcasing a commitment to innovation and customer-centric solutions. The adoption of XR technologies reflects a broader industry trend towards leveraging technology to enhance consumer experiences. Lenskart's embrace of XR technology provides a unique and personalised shopping encounter that goes beyond conventional e-commerce practices, solidifying its position as a forward-thinking brand in the competitive eyewear market. The successful integration of XR into the eyewear shopping journey exemplifies Lenskart's dedication to staying at the forefront of technological advancements, ensuring that customers receive a seamless, immersive, and satisfying shopping experience (Rao et al., 2021).

8.5 Future Directions

XR technology is rapidly expanding across Asia, demonstrating significant growth driven by increasing internet penetration and rising consumer engagement with digital platforms. XR spending in Asia is expected to reach $16.6 billion by 2026, at a projected compound annual growth rate of 42.4% (Communications Today, 2022). In Asia, companies have expanded their operations through XR, particularly in trial-based products such as eyewear. Customers from Asia are experiencing XR on a daily basis on various platforms for their entertainment experiences, like Snapchat and Pokémon Go. Looking ahead, the adoption of XR in Asia is expected to accelerate, presenting opportunities for businesses to leverage XR for enhanced customer engagement, personalised experiences, and innovative product offerings, thereby driving competitive advantage in the evolving digital landscape.

The high initial investment is still a significant barrier; it not only diminishes the scope of pilot projects but also makes returns on investment (ROI) a headache for the management (Akinradewo et al., 2025). To overcome this, firms should partner with local device manufacturers and software developers to co-create lightweight, optimised AR/VR applications tailored to regional languages and cultural contexts. Such collaborations can reduce costs through economies of scale and ensure enhanced end-user understanding by embedding local content. Engagement early in co-designing solutions will mitigate perceived immaturity and build demand confidence (Jha & Singh, 2024).

After investing in XR, Sustained XR adoption hinges on seamlessly blending technology into consumer habits. There are certain framework that can help organisation such as the Virtual Reality Integrated Adoption Framework (VRIAF) which identifies five key determinants: consumer attitudes, perceived enjoyment, ease of use, social influences, and past experiences driving technology uptake (Fares et al., 2024). Enterprises should prioritise streamlined interfaces, gamified feedback loops, collaborative multi-user features, and "try-before-you-buy" pilot programs (e.g., branded XR kiosks) to optimise these dimensions and boost consumer and employee adoption.

Asia's emerging 5G network and edge computing presents a unique opportunity to deliver low-latency, high-fidelity XR experiences (Jha & Singh, 2024). Firms must collaborate with telecom operators for a seamless XR experience over 5G networks.

They should adopt open standards to ensure cross-platform operability. The firms can leverage cloud-based rendering to lower device hardware requirements and broaden user access. Improving interoperability enhances user satisfaction and reduces development overhead by avoiding vendor lock-in. The lack of in-house XR capabilities or organisation amongst various teams slows the initiation and fast expansion of the projects (Noghabaei et al., 2020). Cross-sector partnerships, Academia-Industry Collaborations, and government-industry-Industry Task Forces can bridge this gap by co-develop curricula and apprenticeships focused on XR and launching pilot public-sector XR to derisk private investments, respectively. By embracing localised, cost-effective solutions, seamless user experiences, and strategic partnerships, firms can overcome adoption barriers and unlock XR's full potential across industries.

8.6 Conclusion

This comprehensive review applies a systematic literature review methodology within the PRISMA framework to examine the intricate landscape of XR in marketing, with a specific focus on the Asia Pacific region, with India as a case. The study Analyses 17 pertinent articles from the Scopus database to offer insights into theories, advantages, and disadvantages, as well as the barriers associated with XR adoption in marketing. The review conducts in-depth case studies on two pioneering entities, IKEA and Lenskart, located in India, to provide valuable insights for businesses contemplating XR integration.

The findings highlight the transformative potential of XR technologies across diverse sectors in the Asia Pacific, including healthcare, education, entertainment, corporate training, and real estate. The advantages of XR, including improved customer engagement, collaboration opportunities, and enhanced user experiences, as well as the potential disadvantages and barriers to adoption, such as data security concerns, affordability issues, and technical limitations, are outlined. The case studies exemplify how industry leaders strategically leverage XR to enhance consumer experiences and overcome traditional challenges. This study contributes to theory by enhancing our understanding of XR's role in marketing and consumer behaviour through established frameworks such as co-creation theory, media richness, and flow theory. These insights support the development of new models around user engagement and immersive experience in digital marketing.

The study concludes by proposing a robust future research agenda, emphasising cross-cultural examinations, ethical considerations, advanced methodologies for XR analysis, and sector-specific impacts in healthcare, education, and real estate. Additionally, it advocates for longitudinal studies, research on human–computer interaction, and the development of user-friendly XR content creation tools. Comparative analyses of XR platforms and investigations into XR's impact on consumer trust and brand perception are proposed to further enrich scholarly inquiry in the dynamic domain of XR marketing. This research contributes to the current body of knowledge and provides a roadmap for future exploration and advancements in the transformative realm of extended reality in marketing.

8.6.1 Recommendation of Future Research

The future research agenda in XR marketing encompasses diverse and nuanced avenues for scholarly exploration. Cross-cultural examinations of XR adoption in marketing, particularly in the Asia Pacific nations, pose questions regarding the influence of cultural factors on consumer behaviours and preferences in immersive marketing environments. Ethical considerations in XR marketing present an opportunity for in-depth investigations into privacy concerns, user consent, and the responsible utilisation of immersive technologies. Future research can investigate advanced methodologies for analysing user experiences and behaviour within XR, developing metrics and models to unravel the psychological intricacies of XR interactions. Additionally, exploring the integration of XR in emerging markets involves scrutinising factors such as infrastructure, affordability, and user acceptance, offering insights into the challenges and opportunities for SMEs in these contexts.

The niche industries, including healthcare, education, and real estate, prompt questions about the sector-specific impact of XR on consumer engagement and decision-making processes. Longitudinal studies can be designed to unravel the enduring effects and sustainability of XR marketing initiatives, tracking the evolution of consumer perceptions and brand loyalty over extended periods. Research on human–computer interaction within XR marketing contexts aims to optimise user interfaces, usability, and interaction design principles for enhanced campaign effectiveness. The development of user-friendly XR content creation tools emerged as a critical area, with questions centred on addressing challenges related to content creation in XR, particularly for businesses with limited technical expertise.

Comparative analyses of different XR platforms (AR, VR, MR) are essential, exploring their effectiveness, accessibility, and user acceptance for diverse marketing objectives and target audiences. Lastly, investigating the impact of XR on consumer trust and brand perception raises questions about the factors contributing to trust-building or erosion in XR-based marketing campaigns. These proposed research directions collectively offer a comprehensive roadmap for scholars to contribute to the evolving field of XR marketing, addressing critical gaps and pushing the boundaries of knowledge in this dynamic domain. The study has identified several research questions which are shown in Table 8.4.

Table 8.4 Future research direction for XR (extended reality) research in marketing

Area	Topic	Research questions
Consumer behaviour and psychology	Consumer psychology and behaviour in XR	• How do immersive XR environments impact consumer decision-making? • Which psychological factors (attention, emotion, memory) are most influential in XR?
	Consumer Trust and Ethical Advertising in XR	• What strategies build consumer trust in XR marketing? • How does intrusive advertising in XR affect trust and brand perception?
	Long-term societal effects of XR marketing	• What societal changes result from widespread XR adoption in marketing? • How does XR influence consumer Behaviour and social interactions over time?
Marketing technology and innovation	User experience personalisation in XR	• How can AI adapt XR content in real-time to user preferences? • What is the impact of personalised XR on engagement and conversions?
	Integration of XR with emerging technologies	• How can XR integrate with AI, IoT, and Blockchain to enhance marketing? • What are the benefits and challenges of combining XR with these technologies?
	Standardisation and interoperability	• What industry standards are needed for XR interoperability? • How can companies ensure seamless experiences across XR devices and platforms?
Market accessibility and inclusivity	Accessibility and inclusivity in XR	• How can XR marketing be designed for accessibility? • What design principles improve accessibility for visually, audibly, and physically impaired users?
	Impact of XR on rural and underserved markets	• What barriers exist for XR deployment in rural or underserved areas in Asia Pacific? • How can XR be tailored to meet these markets' needs?

Highlights

- XR integration merges the physical and digital worlds, establishing a transformative framework that Asian companies extensively use to shape the future business landscape.
- The adoption of XR in business operations of Asian companies faces challenges such as high costs, skill deficits, and cultural and privacy issues.
- Companies operating in India—one of the two one billion population country in Asia—are expanding their business through XR across industries such as healthcare, real estate, and tourism. This is evidenced by cases such as IKEA and Lenskart, which have significantly benefited from XR.

References

Akinradewo, O., Hafez, M., Aliu, J., Oke, A., Aigbavboa, C., & Adekunle, S. (2025). Barriers to adopting augmented reality technologies for education and training in the built environment: A developing country context. *Technologies, 13*(2), Article 2. https://doi.org/10.3390/technologies13020062

Alcañiz, M., Bigné, E., & Guixeres, J. (2019). Virtual reality in marketing: A framework, review, and research agenda. *Frontiers in Psychology, 10*. https://doi.org/10.3389/fpsyg.2019.01530

Arya, V., Sambyal, R., Sharma, A., & Dwivedi, Y. K. (2024). Brands are calling your AVATAR in Metaverse—A study to explore XR-based gamification marketing activities & consumer-based brand equity in virtual world. *Journal of Consumer Behaviour, 23*(2), 556–585. https://doi.org/10.1002/cb.2214

Batat, W., & Hammedi, W. (2023). The extended reality technology (ERT) framework for designing customer and service experiences in phygital settings: A service research agenda. *Journal of Service Management, 34*(1), Article 1. Scopus. https://doi.org/10.1108/JOSM-08-2022-0289

Brannon, B., McLean, G., Shah, E., & Mack, R. (2021). Blending the real world and the virtual world: Exploring the role of flow in augmented reality experiences. *Journal of Business Research, 122*, 423–436. https://doi.org/10.1016/j.jbusres.2020.08.041

Brown, R. (2000). Social identity theory: Past achievements, current problems and future challenges. *European Journal of Social Psychology, 30*(6), 745–778. https://doi.org/10.1002/1099-0992(200011/12)30:6%3c745::AID-EJSP24%3e3.0.CO;2-O

Burton, N., & Schlieman, T. (2021). User response to extended reality sponsorship activations on social media: Exploring impressions of GoPro's use of 360° video in marketing. *Journal of Interactive Advertising, 21*(2), 93–107. https://doi.org/10.1080/15252019.2021.1944405

Chadegani, A. A., Salehi, H., Yunus, M. M., Farhadi, H., Fooladi, M., Farhadi, M., & Ebrahim, N. A. (2013). A comparison between two main academic literature collections: Web of science and Scopus databases. arXiv. https://doi.org/10.48550/arXiv.1305.0377

Communications Today. (2022, June 06). AR/VR spending to reach $16.6 billion by 2026. Communications Today. https://www.communicationstoday.co.in/ar-vr-spending-to-reach-16-6-billion-by-2026/

Cummings, J. L., & Teng, B.-S. (2006). The keys to successful knowledge-sharing. *Journal of General Management, 31*(4), 1–18. https://doi.org/10.1177/030630700603100401

de Amorim, I. P., Guerreiro, J., Eloy, S., & Loureiro, S. M. C. (2022). How augmented reality media richness influences consumer behaviour. *International Journal of Consumer Studies, 46*(6), 2351–2366. https://doi.org/10.1111/ijcs.12790

Fares, O. H., Aversa, J., Lee, S. H., & Jacobson, J. (2024). Virtual reality: A review and a new framework for integrated adoption. *International Journal of Consumer Studies, 48*(2), Article e13040. https://doi.org/10.1111/ijcs.13040

Flotyński, J. (2022). Visual aspect-oriented modeling of explorable extended reality environments. *Virtual Reality, 26*(3), 939–961. https://doi.org/10.1007/s10055-021-00601-7

Galvagno, M., & Dalli, D. (2014). Theory of value co-creation: A systematic literature review. *Managing Service Quality, 24*(6), 643–683. https://doi.org/10.1108/MSQ-09-2013-0187

Gong, Z., Nanjappan, V., Soomro, S. A., & Georgiev, G. V. (2021). Virtual brainstorming and creativity: An analysis of measures, avatars, environments, interfaces, and applications. *Proceedings of the Design Society, 1*, 3399–3408. https://doi.org/10.1017/pds.2021.601

Gursoy, D., Lu, L., Nunkoo, R., & Deng, D. (2023). Metaverse in services marketing: An overview and future research directions. *The Service Industries Journal*. https://doi.org/10.1080/02642069.2023.2252750

Ibrahim, F. A., Boying, W., Shafiei, M. W. M., Razak, R. A., & Ishak, N. (2023). Augmented reality (AR) as A promotion tool in influencing housing purchase intention. *Journal of Advanced Research in Applied Sciences and Engineering Technology, 29*(3), Article 3. https://doi.org/10.37934/araset.29.3.251259

Imtiaz, S. A., Krishnaiah, S., Yadav, S. K., Bharath, B., & Ramani, R. V. (2017). Benefits of an android based tablet application in primary screening for eye diseases in a rural population, India. *Journal of Medical Systems, 41*(4), 49. https://doi.org/10.1007/s10916-017-0695-6

Jalo, H., Pirkkalainen, H., Torro, O., Pessot, E., Zangiacomi, A., & Tepljakov, A. (2022). Extended reality technologies in small and medium-sized European industrial companies: Level of awareness, diffusion and enablers of adoption. *Virtual Reality, 26*(4), 1745–1761. https://doi.org/10.1007/s10055-022-00662-2

Jha, A., & Singh, S. R. (2024). Unveiling metaverse applications and challenges in Asia: Opportunities for innovation, marketing, and growth. *Metaverse, 5*(2), 2972. https://doi.org/10.54517/m.v5i2.2972

Jim, J. R., Hosain, Md. T., Mridha, M. F., Kabir, Md. M., & Shin, J. (2023). Toward trustworthy Metaverse: Advancements and challenges. *IEEE Access, 11*, 118318–118347. https://doi.org/10.1109/ACCESS.2023.3326258

Khanra, S., Dhir, A., Kaur, P., & Joseph, R. P. (2021). Factors influencing the adoption postponement of mobile payment services in the hospitality sector during a pandemic. *Journal of Hospitality and Tourism Management, 46*, 26–39. https://doi.org/10.1016/j.jhtm.2020.11.004

Kohli, V., Tripathi, U., Chamola, V., Rout, B. K., & Kanhere, S. S. (2022). A review on virtual reality and augmented reality use-cases of brain computer interface based applications for smart cities. *Microprocessors and Microsystems, 88*, Article 104392. https://doi.org/10.1016/j.micpro.2021.104392

Koohang, A., Nord, J. H., Ooi, K.-B., Tan, G. W.-H., Al-Emran, M., Aw, E. C.-X., Baabdullah, A. M., Buhalis, D., Cham, T.-H., Dennis, C., Dutot, V., Dwivedi, Y. K., Hughes, L., Mogaji, E., Pandey, N., Phau, I., Raman, R., Sharma, A., Sigala, M., Wong, L.-W., etal. (2023). Shaping the Metaverse into reality: A holistic multidisciplinary understanding of opportunities, challenges, and avenues for future investigation. *Journal of Computer Information Systems, 63*(3), 735–765. https://doi.org/10.1080/08874417.2023.2165197

Kumar, A., & Pathak, P. (2018). Synergizing digitization and disruption for service economy: A perspective on digital marketing. *Adhyayan: A Journal of Management Sciences*, 86–99. Business Source Ultimate.

Kumar, H., & Srivastava, R. (2022). Exploring the role of augmented reality in online impulse behaviour. *International Journal of Retail & Distribution Management, 50*(10), 1281–1301. https://doi.org/10.1108/IJRDM-11-2021-0535

Lee, K. (2012). Augmented reality in education and training. *TechTrends, 56*(2), 13–21. https://doi.org/10.1007/s11528-012-0559-3

Martínez-Rojas, M., Pardo-Ferreira, M. del C., & Rubio-Romero, J. C. (2018). Twitter as a tool for the management and analysis of emergency situations: A systematic literature review. *International Journal of Information Management, 43*, 196–208. https://doi.org/10.1016/j.ijinfomgt.2018.07.008

Miles, R. E., & Snow, C. C. (1984). Designing strategic human resources systems. *Organizational Dynamics, 13*(1), 36–52. https://doi.org/10.1016/0090-2616(84)90030-5

Noghabaei, M., Heydarian, A., Balali, V., & Han, K. (2020). Trend analysis on adoption of virtual and augmented reality in the architecture, engineering, and construction industry. *Data, 5*(1), Article 1. https://doi.org/10.3390/data5010026

Ortega-Gras, J.-J., Gómez-Gómez, M.-V., Bueno-Delgado, M.-V., Garrido-Lova, J., & Cañavate-Cruzado, G. (2023). Designing a technological pathway to empower vocational education and training in the circular wood and furniture sector through extended reality. *Electronics, 12*(10), Article 10. https://doi.org/10.3390/electronics12102328

Paul, J., Lim, W. M., O'Cass, A., Hao, A. W., & Bresciani, S. (2021). Scientific procedures and rationales for systematic literature reviews (SPAR-4-SLR). *International Journal of Consumer Studies, 45*(4), O1–O16. https://doi.org/10.1111/ijcs.12695

Rao, P. H. N., Vihari, N. S., & Jabeen, S. S. (2021). Reimagining the fashion retail industry through the implications of COVID-19 in the gulf cooperation council (GCC) countries. *FIIB Business Review, 10*(4), 327–338. https://doi.org/10.1177/23197145211039580

Rauschnabel, P. A., Babin, B. J., tom Dieck, M. C., Krey, N., & Jung, T. (2022). What is augmented reality marketing? Its definition, complexity, and future. *Journal of Business Research, 142*, 1140–1150. https://doi.org/10.1016/j.jbusres.2021.12.084

Shaikh, T. A., Dar, T. R., & Sofi, S. (2022). A data-centric artificial intelligent and extended reality technology in smart healthcare systems. *Social Network Analysis and Mining, 12*(1), 122. https://doi.org/10.1007/s13278-022-00888-7

Stary, C. (2023). Digital process twins as intelligent design technology for engineering Metaverse/XR applications. *Sustainability, 15*(22), Article 22. https://doi.org/10.3390/su152216062

Staw, B. M., & Cummings, L. L. (Eds.). (1984). *Research in organizational behavior* (Vol. 6). JAI Press.

Tan, G.W.-H., Aw, E.C.-X., Cham, T.-H., Ooi, K.-B., Dwivedi, Y. K., Alalwan, A. A., Balakrishnan, J., Chan, H. K., Hew, J.-J., Hughes, L., Jain, V., Lee, V. H., Lin, B., Rana, N. P., & Tan, T. M. (2023). Metaverse in marketing and logistics: The state of the art and the path forward. *Asia Pacific Journal of Marketing and Logistics, 35*(12), 2932–2946. https://doi.org/10.1108/APJML-01-2023-0078

Thomas Alsop. (2023, December 18). "Topic: XR in the United States." Statista. Retrieved October 27, 2024, from https://www.statista.com/topics/7524/extended-reality-xr-ar-vr-and-mr-in-the-us/

Thomas Alsop. (2024, September 2024). "US: AR/VR B2C Market Size 2017–2027." Statista. Retrieved October 28, 2024, from https://www.statista.com/forecasts/1338853/ar-vr-b2c-market-revenue-us

Yadav, S. K., Kumar, A., & Sinha, A. K. (2024). Modeling the critical success factors for continuous usage intention of foodservice mobile apps. *Journal of Foodservice Business Research*, 1–22. https://doi.org/10.1080/15378020.2024.2421051

Zaman, A., Abir, M. R., Mursalin, S., Zaman, A., Abir, M. R., & Mursalin, S. (2024). Extended reality in education and training: Enhancing trustworthiness. *International Journal of Science and Research Archive, 11*(1), Article 1. https://doi.org/10.30574/ijsra.2024.11.1.0206

Nishant Pandey is a student enrolled at the Indian Institute of Management (IIM) Ranchi. He has served as a senior member of the Operations Club & junior executive member of IIM Ranchi's International Relations Committee.

Ankit Kumar Sinha is a Doctoral researcher in the Department of Marketing at the Indian Institute of Management (IIM) Ranchi, Jharkhand, India. He holds a Bachelor of Science from Magadh University and an MBA degree from the Central University of Rajasthan. His research has been published in several academic outlets, including the *Electronic Markets, Journal of Global Marketing, and Journal of Foodservice Business Research* along with proceeding in various national and international conferences.

Anand Kumar is an Assistant Professor at the Institute of Management, Nirma University, Ahmedabad, Gujarat, India. He earned his Ph.D. in Information Systems & Business Analytics from IIM Ranchi and holds an engineering degree from Birla Institute of Technology, India. His research has appeared in prominent journals, including the *Journal of Retailing and Consumer Services and Journal of Foodservice Business Research*, along with numerous international conference proceedings.

Chapter 9
Digital Transformation in Empowering Innovation and Sustainability in the 21st-Century Business Environment in Asia

Clusters of Future Studies: Past and Futures

Chooi Chea Chiam

Abstract Technology is a "two-edge sword" or "two-sides of a coin". Technology and digitalisation transformation are inevitable in the twenty-first century because it has been in every aspect of life, to the extent that twenty-first century is called a time period known as VUCA (Volatility, Uncertainty, Complexity, and Ambiguity), due to its robust, enthusiast and dynamic characteristics. In the fast-paced landscape of the 21st-century business environment, digital transformation has emerged as a catalyst for innovation and sustainability. This paradigm shift leverages advanced technologies to reshape traditional business models, fostering a culture of continuous improvement and adaptability. As organisations navigate the complexities of a globalised economy, embracing digital transformation becomes imperative for staying competitive and relevant. This book chapter discusses the rising importance as well as concerns such as digital transformation for sustainability of business in the challenging twenty-first century Asian business environment, which involves the strategic integration of digital tools and a forward-thinking mindset are essential elements for businesses aiming to thrive in the digital era. The differences in various parameters among Asian countries and the benefits and the challenges on technological innovation were discussed too.

Keywords VUCA · Digital transformation · Sustainability · Business environment · Future of business · Globalised economy

---

C. C. Chiam (✉)
Open University Malaysia, Petaling Jaya, Malaysia
e-mail: chooi_chea@oum.edu.my

M. S. Balaji et al. (eds.), *Future of Business in Asia*,
https://doi.org/10.1007/978-3-031-95940-0_9

9.1 Introduction

Traditional business models have since slowly become obsolete since and there is a need to address the various concerns of the public to ensure the sustainability of their business. Companies in the twenty-first century and beyond have started to adopt green practices, reducing carbon footprints, and embracing sustainability as a core value to meet customer and regulatory demands. Traditional business models have become obsolete and the role of technology has enhanced its presence during the COVID-19 pandemic because it has dramatically underscored the importance of technology across various aspects of society. Digital transformation empowers innovation by providing businesses with the tools and capabilities to harness the vast amounts of data at their disposal. Advanced analytics, artificial intelligence, and machine learning enable companies to gain insights into customer behaviour, market trends, and operational efficiency. These insights, in turn, fuel the innovation engine, allowing businesses to develop new products, services, and processes that meet evolving consumer demands. Furthermore, digital transformation plays a pivotal role in enhancing sustainability efforts. The integration of digital technologies enables businesses to optimise resource utilisation, reduce waste, and minimise their environmental footprint. Automation and smart systems contribute to energy efficiency, while data-driven decision-making helps in identifying areas for improvement in sustainability practices. This not only aligns with global efforts to address climate change but also enhances the brand image by showcasing a commitment to responsible business practices.

Collaboration and communication are also revolutionised through digital transformation, breaking down traditional silos within organisations and facilitating seamless interactions across departments and geographies. Cloud-based technologies and collaborative platforms enable remote work, fostering a more inclusive and diverse workforce. This adaptability is crucial for sustainability, as it allows businesses to navigate disruptions, such as the recent global events that accelerated the shift towards remote work.

According to World Population Review (2024), Asia is the world's largest continent by a wide margin geographically, as well, as the world's most populous continent. This is primarily due to two most populous countries in Asia, China and India, with 1.44 and 1.38 billion people respectively. Asian region consists of approximately 45 to 55 countries. This ambiguity is attributed to the presence of several territories that are considered independent countries by some organisations and governments, but not by others.

This chapter aims to discuss the current and new business environment with its own set of challenges and proposed contingencies for business in the future. This chapter descriptively doing desk-reviews based on past documents, data etc. for the writing.

9.2 Methodology

A conceptual-practice review is undertaken to outline the key concepts and issues in this chapter. Key concepts focused on digital transformation, globalisation, and cultural issues, which were then used as conceptual lenses to explore and understand the issues in practice revolving around digital transformation in empowering innovation and sustainability in the twenty-first century business environment in Asia, a continent that is diversely multi-cultural at the heart of gloablisation. These issues, upon the expert reflection of the authors, include social isolation, digital addiction, technological division, and exposure of personal data.

9.3 Key Concepts

9.3.1 Digital Transformation

Digitisation encompasses the conversion of analog data from offline sources into digital formats, enabling the transition of offline transactions to online platforms. This shift typically does not entail significant changes in process improvement or the creation of new business models within the organisation. Instead, it mainly involves converting various resources and assets into digital forms (Van Veldhoven & Vanthienen, 2023).

The environment of doing business in the twenty-first century is known as a D-VUCAD (Disruption, Volatility, Uncertainty, Complexity, Ambiguity and Diversity) era, which is even more challenging to the Asian countries, which have large and significant influence in doing business in the Asian business context. This reflects a world where change is not just rapid but also deeply disruptive indicating diverse perspectives are essential for navigating complexity. Hence, it is essential to recognise the diversity, trends and shifts across the Asian economies. One of the significant shifts is the digital transformation where technology and innovation has made its presence felt strongly with certain degree of threats to the old business traditional models. The twenty-first century has seen a rapid adoption of technology in the Asian businesses and the trend does not seem to be slowing down or stopping anytime in the near future.

New method of promoting products and making payments were introduced such as e-commerce, mobile payments, and digital platforms have transformed the way transactions are conducted. Several large nations have widely adopted cashless payment approach for everyday transactions. For example, China makes the headlines, which the country serves as a prominent example and is on the verge of achieving a cashless society. China initiated its transition towards a cashless society in 2017, with over 75% of its population favoring digital payments over traditional cash transactions. The younger demographic in China spearheaded the adoption of digital payment methods, subsequently influencing older generations to embrace these practices. This swift and comprehensive integration of digital payments into mainstream transactions has facilitated the Chinese government's efforts in this regard. (Travel Weekly Asia, 2023). Before the end of year 2022, the increasing of prevalence of cashless payments in China is causing unease among banking regulators and users as the influence of cashless payment in the urban areas as well as tourists encounter difficulties in providing change for cash or, at times, decline cash transactions altogether. Tourists find that total cashless in China can be a hassle due to fact that they not have local bank account to the local platform such as Wechat pay, Alipay etc. The rising trend of "cashlessness" in Chinese urban centers poses a risk of revealing latent economic instability.

Mobile payments are creating divisions between different age groups and accentuating disparities between the affluent urban middle class and those who have not benefited from the economic boom. Traditionally, foreigners had to establish a Chinese bank account to gain access to China's mobile payment networks. Consequently, the country's mobile payment system became largely inaccessible to overseas tourists, especially considering that mobile apps became the predominant mode of payment within the nation. Therefore, international travelers in China have encountered increasing difficulties in making payments for goods and services, given that numerous shops and restaurants now exclusively accept mobile payments, excluding credit cards or cash. However, since, China has acknowledged the problem faced by foreign tourists. China since has introduced the acceptance of international credit or debit cards such as Visa via WeChat Pay from mid-July of year 2023 for foreign tourists to China to tackle the problem and to move the country to digital reformation on the next level.

According to Linkedin (2023), in the backdrop of the onset of the 4th industrial revolution and the global COVID-19 pandemic, non-cash transactions have significantly transformed the lifestyles of people worldwide, including those in Asian countries. Examining the prevailing trend of cashless payments in Asia becomes crucial for discerning the variations among nations. Owing to its vast population, China's adoption of cashless payments may not match that of South Korea; however, China remains the regional frontrunner in the field of e-commerce. Nevertheless, Thailand, Japan, and Vietnam trail behind as the top three slow adopters of cashless payments in Asia, despite the region spearheading the global shift towards digital transactions. This may be because Japan, a country with more mature economy and an aging population, cash continues to be prevalent. This is attributed partly to the extensive Automated Teller Machine (ATM) network and comparatively high credit card fees, discouraging small merchants from readily embracing cashless payments.

Firms are currently leveraging advanced analytics and machine learning to predict customer demands, streamline supply chains, and offer personalised experiences (Attaran, 2020). This advancement has propelled us into an era where organisations are not just embracing digital tools but are fundamentally reshaping their operations, competition strategies, and value creation processes (Omol et al., 2023). The digital transformation journey has emerged as a pivotal component of organisational strategy, exerting influence across all facets of the contemporary business environment.

The digital transformation unfolding in South and Southeast Asia is presenting many opportunities for many, particularly the younger generations. Amidst the COVID-19 pandemic, digital connectivity in Asia proved indispensable in surmounting the challenges posed by traditional trade methods. As per the findings of the International Monetary Fund (IMF) in 2018, there exists a correlation between gross domestic product (GDP) per capita and the utilisation of the digitalisation index (UDI) in 2016. Specifically, the higher the GDP per capita, the greater the usage of digitalisation index. Among the eleven countries in Southeast Asia (SEA), namely Brunei, Cambodia, Indonesia, Laos, Malaysia, Myanmar, Philippines, Singapore, Testimo Leslie, Thailand and Vietnam, Singapore boasts the highest UDI, succeeded by Malaysia, Thailand, the Philippines, and Vietnam. Conversely, Myanmar exhibits the lowest UDI, while Indonesia, Cambodia, and Laos fall between these two extremes.

9.3.2 Globalisation

Globalisation is characterised as a phenomenon where interconnectedness and interdependence of economies, cultures, and societies across the globe has increased, has had a profound impact on Asian countries. As the world becomes increasingly integrated through trade, investment, technology, and cultural exchange, Asian nations have found themselves at the forefront of this transformative process. The effects of globalisation are palpable, shaping the economic, social, and cultural landscapes of these diverse and dynamic Asian countries Asia has played a central role in global trade and commerce, serving as a crossroads for cultural exchange and economic activity. However, in recent decades, the pace and scale of globalisation have accelerated, propelled by advancements in technology, transportation, and communication. This rapid integration into the global economy has brought both opportunities and challenges for Asian nations, transforming their economies, societies, and ways of life.

Globalisation has opened up new markets and opportunities for Asian countries, fueling rapid economic growth and development economically. Countries such as China, India, and South Korea have emerged as major players on the global stage, leveraging their abundant labor resources, technological prowess, and strategic location to become key drivers of global economic growth. The rise of multinational corporations and the expansion of global supply chains have facilitated the flow of goods, services, and capital across borders, leading to increased trade and investment in the region. However, alongside the benefits of globalisation, Asian countries have also grappled with its challenges. Economic liberalization and market reforms have led to rising income inequality, environmental degradation, and social dislocation in many parts of the region. Moreover, the increasing interconnectedness of Asian economies has made them more vulnerable to external shocks and crises, as evidenced by the Asian financial crisis of the late 1990s and the more recent global economic downturn triggered by the COVID-19 pandemic.

In addition to its economic impact, globalisation has also had profound social and cultural implications for Asian societies. The spread of foreign culture, the rise of social media and digital technologies, and the influx of foreign ideas and influences have reshaped traditional norms and values, leading to debates about identity, cultural authenticity, and social cohesion. Moreover, the mobility of people across borders has given rise to multicultural societies, fostering greater diversity and cosmopolitanism in many Asian countries. Globalisation has emerged as a defining feature of the contemporary world, profoundly shaping the economic, social, and cultural dynamics of Asian countries. While presenting new opportunities for growth and development, it also poses significant challenges and uncertainties for policymakers, businesses, and individuals alike. As Asian nations navigate the complexities of globalisation, they must strive to harness its benefits while mitigating its adverse effects, in order to build inclusive, sustainable, and resilient societies for the future.

Globalisation has firmly entrenched itself within the economic fabric of Asian countries. This is unmistakably illustrated by the proliferation of multinational corporations, the intricate networks of global supply chains, and the surge in cross-border investments. The surge in international trade and economic collaborations has ushered in a new era of both opportunities and challenges for businesses operating within the region.

Entrepreneurship and startup initiatives have experienced a notable upswing across numerous Asian nations. Governments in the region have acknowledged the pivotal role of fostering innovation and entrepreneurial spirit, thus implementing supportive policies, establishing incubators, and initiating funding programs to bolster these endeavors. The twenty-first century has marked a pronounced redistribution of economic power towards Asia. Notably, China has risen as a dominant global economic force, influencing trade dynamics, investment patterns, and corporate strategies on a worldwide scale.

Economic reforms have been undertaken by numerous Asian countries to bolster their business environments. These reforms have entailed streamlining regulatory frameworks, implementing tax reforms, and enhancing infrastructure to allure foreign investments and catalyse domestic business expansion. Changes in demographics and escalating incomes have instigated transformations in consumer behaviour. This has precipitated a burgeoning demand for superior products and services, compelling businesses to adjust their strategies to align with these evolving market dynamics. Alterations in demographics and escalating incomes have instigated shifts in consumer behaviour. This has resulted in a mounting demand for high-quality products and services, prompting businesses to adapt in order to cater to these evolving market dynamics.

Sustainability and Corporate Social Responsibility (CSR) have garnered increasing attention and emphasis within Asian businesses. Companies are acknowledging the significance of environmental and social responsibility in meeting the expectations of both consumers and investors. In the global business landscape, effective cross-cultural management has become indispensable as businesses expand their operations internationally. The ability to comprehend diverse cultural nuances and adjust management approaches accordingly has emerged as a vital component for achieving success in Asian markets. Asian governments have proactively championed business-friendly policies and initiatives aimed at attracting foreign investments and fostering economic expansion. These encompass a spectrum of measures such as tax incentives, infrastructure enhancements, and endeavors to streamline bureaucratic processes.

Amidst the vast opportunities present in the Asian market, businesses encounter a myriad of challenges including regulatory intricacies, cultural disparities, and fierce competition. Success hinges on adaptability and a profound comprehension of the local business landscape. In essence, the business landscape of Asian countries in the twenty-first century is characterized by technological progress, globalisation, economic restructuring, and a dynamic reconfiguration of economic dominance, presenting a spectrum of challenges and opportunities for enterprises operating within the region. While it's challenging to generalise about doing business across entire continents, there are some notable differences between doing business in Asian countries and other parts of the world. These differences can be attributed to cultural, economic, regulatory, and historical factors.

9.3.3 Cultural Differences

There could be differences in the communication styles among different cultures and countries. Asian cultures often place a high value on indirect communication and maintaining harmony. This can contrast with more direct communication styles prevalent in some Western countries. Asian cultures tend to have a strong emphasis on hierarchy and respect for authority. Understanding and navigating these hierarchies is crucial for successful business interactions. The legal systems in Asian countries, influenced by diverse historical and cultural factors, may differ significantly from those in other regions. Understanding local laws and regulations is essential for compliance and risk management. Several Asian countries may have more bureaucratic processes compared to others, which can affect the ease of doing business.

Consumer behaviour can vary widely across regions. Understanding local preferences, purchasing power, and cultural nuances is crucial for marketing and product/service customisation. The level of market maturity, competition, and saturation can differ greatly between Asian countries and other regions, influencing market entry strategies. In some Asian countries, there may be a higher degree of government involvement in business affairs, ranging from regulatory controls to state-owned enterprises. This contrasts with more market-driven economies in certain Western countries. Building strong personal and business relationships is often emphasised in Asian business cultures. This is reflected in the importance of trust and long-term commitments in business partnerships. Networking practices and relationship-building strategies can differ, with some regions placing a higher emphasis on formal business networks and introductions. Differences workforce dynamics could occur from the perspective of labour practices and employment regulations can vary widely.

Understanding local labor markets and employment laws is crucial for managing human resources effectively. The expectations of the work-life balance can be very different and practice. Expectations regarding work hours and work-life balance can differ between Asian countries and other regions.

The level of technology adoption and digital infrastructure can vary. Some Asian countries are at the forefront of technological advancements, while others may be in earlier stages of development. Risk tolerance and attitudes toward uncertainty can differ, influencing business strategies and decision-making. Responses to crises, such as economic downturns or geopolitical tensions, may vary based on regional perspectives and experiences. It's important to note that these are general observations, and there is significant diversity within both Asian countries and other regions. Each country has its unique business environment shaped by historical, cultural, and economic factors. Successful international businesses often adapt their strategies to accommodate these differences and build strong local partnerships. Certainly, cultural differences play a significant role in shaping business practices. Table 9.1 some examples of cultural distinctions between doing business in Asian and Western countries:

It's crucial to approach these cultural differences with sensitivity and an open mind. Successful international business ventures often involve a deep understanding and respect for the cultural nuances of the regions involved. As it is, countries in the Asian region could have different practices, legislation and customary in doing business for small medium enterprises (SMEs) and micro small medium enterprises (MSMEs). As illustrated in Table 9.2 by Gano-an and Chiam (2021), there are 11 parameters that were studied and compared between two Asian countries, namely, Malaysia and Philippines.

9.4 Key Issues

Although technology and social media present a plethora of benefits, it's imperative to acknowledge potential drawbacks, including privacy breaches, the spread of misinformation, and their effects on mental well-being. Therefore, maintaining a balance between leveraging the positive aspects and practising responsible usage is essential for maximising the advantages offered by these tools.

9.4.1 Social Isolation

One prominent drawback of technology is the phenomenon of technological social isolation, which leads to a decline in face-to-face interaction among individuals. Despite technological advancements enabling better connectivity through various social platforms, the quality of online interaction differs significantly from physical social interactions. Face-to-face communication plays a vital role in fostering deep,

Table 9.1 Cultural differences between Asian and Western countries in doing business

Asian countries	Western countries
Communication style	
Value indirect communication. Messages may be conveyed subtly, and there's often a preference for non-verbal cues and gestures	Generally, favor more direct communication. Clear and explicit verbal communication is often expected in business interactions
Hierarchy and authority	
Often more pronounced. Respect for authority and age is crucial, and decisions may be made by senior members of the organisation	While hierarchies exist in Western organisations, there tends to be a more egalitarian approach. Decision-making can involve collaboration, and individuals may feel comfortable challenging authority
Relationship building	
Building strong personal relationships is a cornerstone of business in many Asian cultures. Business decisions often rely on trust, and there's an emphasis on long-term partnerships	While relationships are important in Western business, the focus may be more on the professional aspects, and business dealings can be more transactional
Work ethic and time management	
Many Asian cultures place a strong emphasis on hard work and dedication. Longer working hours are often the norm, and punctuality is highly valued	While relationships are important in Western business, the focus may be more on the professional aspects, and business dealings can be more transactional
Decision-Making process	
Decision-making can be a more consensus-driven process in Asian cultures. It may involve consultation with various stakeholders before reaching a conclusion	Decision-making in Western cultures can be more individualistic, with decisions often made by a designated authority figure or a smaller group of leaders
Negotiation styles	
Negotiations in Asia may involve a more gradual process, with an emphasis on building relationships before getting into the details of a deal. Face-saving is crucial	Negotiations in Western cultures can be more direct, focused on the facts and figures. The emphasis may be on achieving a mutually beneficial agreement rather than necessarily preserving face
Gift-Giving and etiquette	
Gift-giving is often a common practice in Asian business culture, symbolising respect and appreciation. There are often specific rules regarding the type and timing of gifts	While gift-giving is not uncommon in Western business, it may be less formalised and is often not as integral to business interactions as in many Asian cultures
Personal space and physical contact	
Personal space is often smaller, and physical contact may be more acceptable, reflecting a closer interpersonal distance	Personal space is typically larger, and physical contact is more reserved, with handshakes being the common form of greeting

Source Adapted Pattanayak (2024), Fly homes

Table 9.2 Parameters for ease of doing business

Indicator set	Areas measured
Starting a business	This pertains to the procedures/steps, time, cost and paid-in minimum capital to start a limited liability venture
Dealing with construction permits	This relates to the processes, time and cost to accomplish all permits required to build a warehouse and the quality control and safety standards in the construction permitting system
Getting electricity	This concerns the cost, process, and time in getting connected to the electrical grid, the reliability of the power supply and the transparency of tariffs
Registering property	The cost, length of time, and procedures to transfer a property and the quality of the land administration system
Getting credit	The country's credit information systems and movable collateral laws
Protecting minority investors	This tackles the rights of minority shareholders pertaining to business governance and related-party transactions
Paying taxes	The tax rates, fees, and time necessary for the business to fulfill the tax obligations and other related processes
Trading across borders	Cost and time necessary to export goods of comparative advantage and import auto parts
Enforcing contracts	This concerns the cost and time to resolve business disputes and the features/ nature of the judicial processes of a country
Resolving insolvency	Cost, time and rate of recovery from insolvency and standards of the legal structure for resolving insolvency
Labor market regulation	The employment regulations and practices of a country including the quality of job

Source World Bank Group (2019), Gano-an and Chea (2021)

enduring, and meaningful relationships. However, online interactions often prompt people to replace physical communication with virtual conversations, resulting in the illusion of vast online social circles devoid of genuine, tangible, and physical connections. The social media aid in alleviating the feeling of "loneliness" to some, however, over-reliance of social media tends to have negative effect because a study found that intrapersonal reasons such as wanting to escape daily problems or simply pass the time are strong predictors of problematic social media use (Schivinski et al, 2020).

The absence of in-person social interactions has contributed to an increase in various mental health concerns, such as anxiety, depression, and low self-esteem, addiction to technology gadgets etc. Continuous exposure to idealised portrayals of others' lives on social media platforms can foster feelings of inadequacy and worthlessness. Additionally, the addictive nature of social media, fueled by algorithms aimed at maintaining user engagement, can worsen these mental health issues, perpetuating a cycle of negative emotions. As the technology advances, the issue of personal privacy arises where people post and share their daily life matters to the public. Apart from that, the ability and capacity of the algorithm has increased tremendously in storing, collecting, and analysing vast amounts of personal data. The breach of data and privacy in the digital age raises serious concern. The seriousness of the concern has reached a level where there is misuse of personal data, where people steal one's identity and details in applying for loans as well as withdrawing deposits from the bank under one's stolen identity. There is a predator hiding in the web lurking and targeting one's details such as flight tickets, family members' photos, car plate numbers, house numbers etc.

9.4.2 Digital Addiction

The convenience and entertainment offered by technology have led to a rise in digital addiction, characterised by a constant need to check notifications, messages, emails, and social media feeds. This compulsive behaviour can have negative effects on productivity, sleep patterns, and overall well-being. While technology has undoubtedly increased efficiency and accuracy in tasks, as well as brought convenience to people's lives, its impact on productivity can also lead to job displacement. Certain tasks once performed by humans are now carried out by machines, resulting in economic challenges such as reduced job opportunities and income inequality. As a result of these changes, there is a growing need for upskilling and training in 21st-century skills to adapt to the evolving job market and technological landscape.

The capacity of technology, particularly through social media, to rapidly disseminate information to a broad audience is undoubtedly one of its strengths. However, there is a flip side to this coin, as false news, narratives, and misinformation can be propagated with malicious intent. These misleading narratives have the potential to sway public opinion, shape perceptions, and even influence political events. Netizens' reactions and comments can sometimes sway issues or verdicts, impacting events significantly. Moreover, many individuals assume the roles of "social moralists" or "moral police," dictating how others should behave, dress, or speak. This phenomenon has led to the proliferation of cyberbullying, which can have dire consequences such as suicide or public displays of erratic behaviour. Given that many lacks the skills to discern the authenticity of information, there is an urgent need for effective mechanisms to verify information on social media. The failure to address this issue poses a serious threat to democratic processes and public discourse.

Other than that, children's social skills, especially those below the age of twelve may be compromised as they increasingly rely on social media for interaction, often at the expense of face-to-face communication. While they may excel in making online friends, they may lack proficiency in engaging in everyday conversations. Additionally, the pervasive influence of trending media sensations can divert children's attention from subjects like history and politics, altering their perspectives and behaviour. Refer to Fig. 9.1 for the challenges of technology innovation in Asian countries which have been discussed earlier.

In recent years, internet usage has increased significantly, hence, excessive use have also become more evident Internet addiction disorder is a complex issue that

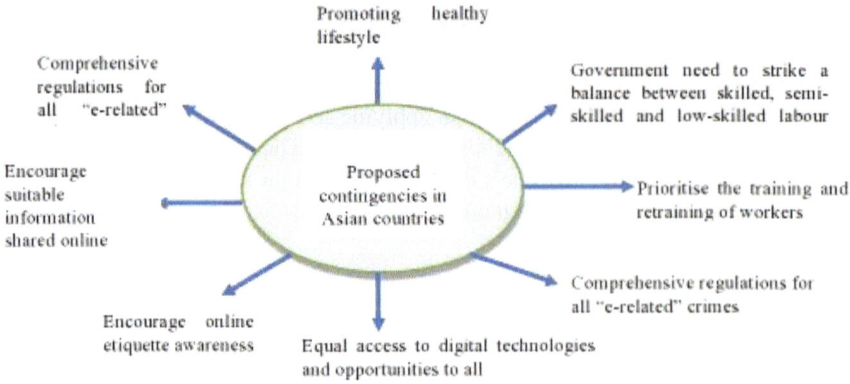

Fig. 9.1 Challenges of technological innovation in Asian countries

spans multiple disciplines including medicine, computer science, sociology, law, ethics, and psychology each exploring it from different perspectives. The rising volume of research on internet addiction highlights it as a psychosocial disorder, commonly characterized by symptoms such as increased tolerance, withdrawal effects, mood disturbances, and difficulties in social relationships (Alavi et al., 2011).

9.4.3 *Technological Division*

Another downside of technology is the phenomenon known as technological division, which results in unequal access to technology, leading to a digital divide. Discrepancies in internet access, availability of digital devices, and technological education can amplify pre-existing social and economic disparities. This issue is particularly prevalent in impoverished and rural areas, where the financial burden of maintaining consistent high-speed internet connectivity can be overwhelming for residents. Additionally, internet infrastructure in rural regions is often inadequate, resulting in weak and unreliable connectivity. Bullying has evolved beyond physical or verbal forms to include cyberbullying, where netizens engage in harsh comments and words to enforce their notion of "justice." Social media platforms also serve as channels for spreading malicious and false news, aiming to traumatise and manipulate public perceptions. Many cases shared with the public are often judged based on one-sided narratives, fueling public sentiments. However, social media can also provide an alternative platform for those seeking justice, offering hope to victims of injustice. On the flip side, it's also used for online harassment, threats, and humiliation, causing long-term emotional and psychological effects on victims.

The term "digital divide" refers to inequalities in the use of digital technologies where there is a gap between those who can use technology effectively and those who cannot. According to Gaillard (2001), the digital divide represents the boundary between those who have access to computers, the necessary skills, and internet usage, and those who lack both access to technology and the internet.

9.4.4 *Exposure of Personal Data*

Moreover, there's been a surge in people flaunting their lifestyles on social media without fully grasping the potential risks, including privacy breaches, identity theft, and misuse of personal data. This trend is exacerbated by the increasing number of scamming cases, where predators pose as individuals' friends or romantic partners. Individuals with malicious intentions often exploit social media platforms to disseminate false information and fake news, leading to societal harm. Moreover, excessive use of social media can result in health issues such as anxiety, depression, and low self-esteem, as individuals may struggle with maintaining different personas in real life and their virtual presence.

Digital technologies enable companies to collect and analyse vast amounts of data, which they can leverage to boost profitability (data monetization) or enhance performance across their broader business networks (data sharing). Data monetization involves using data for direct or indirect economic gain. One approach, known as data wrapping, uses data-driven insights to develop new products and services tailored to customer needs. For instance, AirAsia, an aviation company could track how often/regular an account checking the price of the flight to certain destinations at certain flying dates with other variables in place and the price varies each time surfing/checking using the same particular account or even IP address of the technology tools such as mobile phones, laptops etc., prompting to push the customer to make the purchase earlier. This goes the same with the side purchases such as additional baggage check-in, meals etc.

9.5 Discussion of Implications

9.5.1 *Proposed Contingencies for Technological Innovation Challenges in Asian Countries*

There is an urgent need for comprehensive legislation covering cybercrimes is particularly urgent due to the region's rapid technological advancement and increasing reliance on digital infrastructure in Asian countries. One of the primary reasons for comprehensive cybercrime laws in Asian countries is the exponential growth of internet connectivity and digital services across the region as well as more individuals and businesses go online, the potential for cybercrimes such as hacking, online fraud, and identity theft escalates. People and organisations are left vulnerable to exploitation by cybercriminals who exploit loopholes in existing laws or operate beyond the reach of outdated regulations without adequate legal safeguards. Interconnected nature of global cyberspace means that cyber threats transcend national borders. Asian countries often find themselves targeted by cyberattacks originating from both domestic and foreign actors. Comprehensive cybercrime legislation is essential to facilitate international cooperation and coordination in investigating and prosecuting cybercriminals, as well as to establish mechanisms for information sharing and mutual assistance with other countries.

9.5.2 Develop Digitally Adept Workforce

The development of a regional digital market hinges on having a digitally adept workforce. Therefore, Asian countries must narrow the digital gap by bolstering connectivity, notably through a broadband revolution across the region. Additionally, it is imperative to equip the workforce with essential digital skills and cultivate a mindset conducive to embracing digitalisation. To foster digital inclusion, Asian countries should prioritise education for both the younger and older generations, addressing the surging. demand for digital skills and promoting lifelong learning (Bush & Najar, 2022; Chao, 2022; International Monetary Fund, 2018; Phanthavong, 2022). Moreover, companies must concentrate on fostering digital cultures within their organisations (Laker, 2021). Governments should prioritise the training and retraining of workers, particularly seniors, who may be adversely affected by the rapid advancements in digital technologies. Many older workers harbor concerns about potential displacement by machines and robots, leading to resistance to change, which is a significant barrier to digital adoption. Hence, training programs should aim to instill a willingness among older workers to embrace digital transformation. However, it's important to note that proficiency in hard skills alone is insufficient for workers to thrive in their roles or careers, as both technical and soft skills hold equal importance in the digital economy (The Economist Intelligence Unit, 2019).

9.5.3 Enhancing Cybersecurity Capabilities

Furthermore, it is imperative for Asian countries to collaborate and enhance their cybersecurity capabilities on a regional scale. At the national level, there is a pressing need to enhance trust and security within online platforms and marketplaces, as incidents related to security and privacy often discourage consumers from engaging online (Di, 2018; Ha, 2017; Ha & McGregor, 2013). In order to address cyber incidents, Asian countries must establish a secure digital marketplace. This can be achieved through the exchange of critical information, the creation of protocols and manuals for handling hacking incidents, enhanced coordination among national cybersecurity agencies, and the establishment of a regional framework for dispute resolution (Global-is-Asian Staff, 2018). Civil society organisations have the

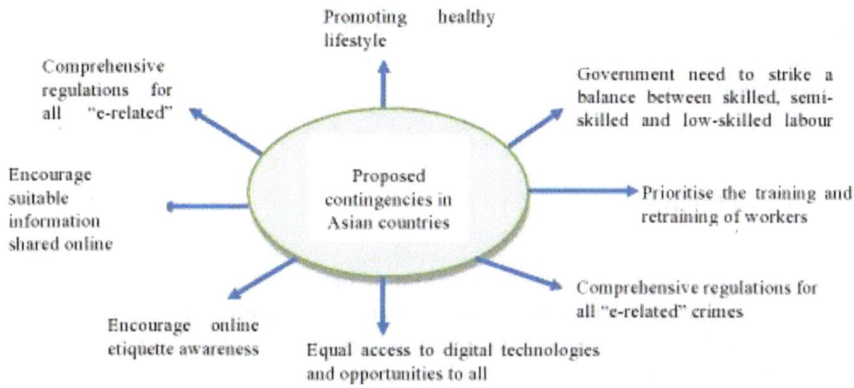

Fig. 9.2 Proposed contingencies for the challenges of technological innovation in Asian countries

capacity to advocate for cybersecurity awareness within communities, promote the adoption of cyber resilience management practices, engage in pertinent cybersecurity projects, facilitate cybersecurity capacity-building initiatives, and forge partnerships to enhance overall cybersecurity efforts (Un et al., 2021). Asian countries must foster both intra- and inter-digital transformation to bolster their national and regional economies. Achieving this requires bridging the gaps between "Industry 4.0," "Technology 4.0," "Education/Training 4.0," and "Policy 4.0" at the local, national, and regional levels (OECD, 2017). Additionally, these countries should prioritise the development and implementation of fair competition policies, along with substantial investments in digital education (human capital) and digital infrastructure. Such initiatives can stimulate digital innovation, promote inclusiveness, and contribute to establishing a secure and integrated digitally connected region (Mulyani, 2017, p. 31). Refer to Fig. 9.2 with proposed contingencies for the technological innovation challenges in Asian countries based on Fig. 9.1 for its challenges of technological innovation in Asian countries.

9.6 Future Directions

Businesses in Asia need to leverage on the technology as one of the key future directions is the acceleration of digital transformation because advancements in artificial intelligence (AI), blockchain, and big data analytics could enhance efficiency and competitiveness. Other than that, environmental concerns are also increasingly influencing business strategies in Asia where governments are implementing stricter regulations on carbon emissions, waste management, and renewable energy adoption. Hence, firms investing in green technologies, circular economies, and sustainable supply chains will gain a competitive edge. Another essential aspect is understanding cultural differences as globalisation connects Asian markets to the world. The future of business in Asia will be shaped by technology, sustainability, and workforce transformation. Firms that embrace these trends will thrive in the dynamic 21st-century business landscape.

9.7 Conclusion

Digital transformation stands at the forefront of shaping the modern business landscape, serving as a catalyst for innovation and sustainability in the twenty-first century Asian business environment. This transformative process has revolutionised how organisations operate, interact, and innovate, fundamentally reshaping the way businesses navigate the ever-evolving digital age. By integrating advanced technologies and digital strategies, businesses are empowered to adapt, thrive, and lead in an increasingly competitive and dynamic environment. At the heart of digital transformation lies the integration of cutting-edge technologies, such as artificial intelligence, big data analytics, cloud computing, and the Internet of Things (IoT). These technologies have enabled businesses to harness vast amounts of data, derive actionable insights, and drive informed decision-making. By leveraging data-driven approaches, organisations can identify market trends, anticipate customer needs, and tailor products and services to meet evolving demands.

Furthermore, digital transformation has revolutionised traditional business models, fostering greater agility, flexibility, and responsiveness. Through the adoption of digital platforms and agile methodologies, organisations can streamline processes, optimise workflows, and accelerate time-to-market. This agility enables businesses to swiftly adapt to changing market conditions, seize emerging opportunities, and stay ahead of the competition. Moreover, digital transformation has democratized innovation, empowering businesses of all sizes to innovate and disrupt industries. By providing access to cutting-edge technologies and resources, digital platforms have leveled the playing field, allowing startups and small businesses to compete with industry giants. This democratisation of innovation has fueled a surge in entrepreneurial activity, driving a wave of disruptive innovations across various sectors. In addition to driving innovation, digital transformation has emerged as a powerful force for sustainability in the business world. By leveraging digital technologies, businesses can enhance sustainability efforts, reduce environmental impact, and promote responsible business practices. For instance, IoT-enabled sensors and smart devices can optimise resource utilisation, minimise waste, and improve energy efficiency. Similarly, cloud computing and remote collaboration tools can facilitate remote work, reducing the carbon footprint associated with commuting and office space.

Besides, digital transformation has enabled businesses to adopt sustainable business models and practices, aligning profitability with environmental and social responsibility. From renewable energy initiatives to supply chain transparency efforts, organisations are leveraging digital technologies to embed sustainability into their core business operations. This integration of sustainability into business strategies not only benefits the planet but also enhances brand reputation, attracts environmentally conscious consumers, and drives long-term value creation. However, while digital transformation offers immense opportunities for innovation and sustainability, it also presents unique challenges and risks. One such challenge is the need for organisations to navigate the complexities of digital disruption and ensure that their workforce is equipped with the necessary skills and competencies to thrive in the digital age. Additionally, concerns related to data privacy, cybersecurity, and ethical implications of artificial intelligence (AI) pose significant challenges that must be addressed. Moreover, the digital divide remains a pressing issue, with disparities in access to technology and digital literacy exacerbating social and economic inequalities. Bridging this digital divide requires concerted efforts from governments, businesses, and civil society to ensure that all individuals have equal access to digital technologies and opportunities.

Digital transformation is reshaping the business landscape, empowering organisations to innovate, compete, and thrive in the twenty-first century Asian business environment. By embracing advanced technologies, businesses can drive innovation, enhance efficiency, and foster sustainability across their operations. However, realising the full potential of digital transformation requires organisations to address challenges, bridge divides, and embrace responsible and inclusive approaches to digitalisation. By doing so, businesses can unlock new opportunities, create value, and contribute to a more sustainable and prosperous future for all.

Highlights

- Digital transformation is revolutionising business in Asia by driving innovation and enhancing sustainability but posed a great risk of security simultaneously, such as great exposure of risk in identity theft, loss of social skills, and digital-addiction, among others
- Firms in Asian need to leverage on technology to be competitive with the other firms in the world where contingencies such as better-rounded laws and regulations on cybersecurity are in place in the country to bring the stability of technology advancement in Asian countries.
- Companies in Asia regardless of whether they originated from Asia or originating elsewhere but operating in Asia need to embrace, albeit carefully (due to inherent risks) rather than cautiously (due to inherent opportunities), digital solutions to forge and sustain a competitive edge in the rapidly evolving VUCA business landscape because emerging technologies are improving efficiency, reducing environmental impact, and fostering new business models.

References

Alavi, S. S., Maracy, M. R., Jannatifard, F., & Eslami. M. (2011). The effect of psychiatric symptoms on the internet addiction disorder in Isfahan's University students. *Journal of Research in Medical Science, 16*(6), 793–800. PMID: 22091309; PMCID: PMC3214398.

Attaran, M. (2020). Digital technology enablers and their implications for supply chain management. *Supply Chain Forum: An International Journal, 21*(3), 158–172, Taylor & Francis. https://doi.org/10.1080/16258312.2020.1751568

Bush, R., & Najar, H. (2022). Digital skills for inclusive growth: Reaching the hard-to-read in Southeast Asia. *The ASEAN, 23*, 24–25.

Chao, R. Y., Jr. (2022). ASEAN community-building through digital transformation of education systems. *The ASEAN, 23*, 13–15.

Di, D. (2018). World Economic Forum initiative pledges to equip 20 million ASEAN workers with digital skills by 2020. World Economic Forum. Available from: https://www.weforum.org/press/2018/11/world-economic-forum-initiative-pledges-to-equip-20-million-asean-workers-withdigital-skills-by-2020/. Accessed 22 March 2024.

Gaillard, F. D. (2001). Understanding the digital divide as it relates to electronic commerce. Unpublished doctoral dissertation, Old Dominion University.

Gano-an, J. C., & Chea. C. C. (2021). Ease of doing business in the Philippines and Malaysia: Prospects for development and regional growth. *American Journal of Business and Society, 6*(2), 49–56. Public Science Framework.

Global-is-Asian Staff. (2018). *Building the digital economy in ASEAN*. Lee Kuan Yew School of Public Policy, National University of Singapore. Available from: https://lkyspp.nus.edu.sg/gia/article/building-the-digital-economy-in-asean. Accessed 22 March 2024.

Ha, H. (2017). Stakeholders' views on self-regulation to protect consumers in E-retailing. *Journal of Electronic Commerce in Organizations, 15*(3), 83–103.

Ha, H., & McGregor, S. L. T. (2013). Role of consumer associations in the governance of E-commerce consumer protection. *Journal of Internet Commerce, 12*(1), 1–25. https://doi.org/10.1080/15332861.2013.763691

International Monetary Fund. (2018). Asia and Pacific—Asia at the forefront: Growth challenges for the next decade and beyond [Video], October 10. Available from: https://meetings.imf.org/en/2018/Annual/Schedule/2018/10/10/IMF-seminar-asia-at-the-forefront. Accessed 22 March 2024.

Laker, B. (2021). Five creative ways to lead a digital transformation. Forbes, available from: https://www.forbes.com/sites/benjaminlaker/2021/10/07/five-creative-ways-to-lead-a-digitaltransformation/?sh57f744f036234. Accessed 22 March 2024.

Linkedin. (2023). *Cashless payment trend in Asia.* Retrieved from https://www.linkedin.com/pulse/cashless-payment-trend-asia-henovn. Assessed 2 February 024.

Mulyani, I. (2017). ASEAN in the digital era: Opportunities and challenges. *CIMB Southeast Asia research Sdn Bhd, advancing ASEAN in the digital age book* (pp. 29–31). CIMB Southeast Asia Research Sdn Bhd.

OECD. (2017). *Opportunities and policy challenges of digitalisation in Southeast Asia.* OECD.

Omol, E., Mburu, L., & Abuonji, P. (2023). Digital maturity action fields for SMEs in developing economies. *Journal of Environmental Science, Computer Science, and Engineering & Technology, 12*(3). https://doi.org/10.24214/jecet.B.12.3.10114

Pattanayak, A. (2024). *Asian culture vs western culture: Differences & similarities. flyhomes.* https://fly.homes/blog/culture/asian-culture-vs-western-culture/. Accessed April 2024.

Phanthavong, E. (2022). Driving inclusive and sustainable digital transformation in ASEAN. *The ASEAN, 23*, 5–7.

Schivinski, B., Brzozowska-Woś, M., Stansbury, E., Satel, J., Montag, C., & Pontes, H. M. (2020). Exploring the role of social media use motives, psychological well-being, self-esteem, and affect in problematic social media use. *Frontiers in Psychology*, 11, https://doi.org/10.3389/fpsyg.2020.617140

The Economist Intelligence Unit. (2019). *The digitalisation of commerce in Asia: Expanding the rewards and addressing the risks.* The Economist Intelligence Unit

Travel Weekly Asia. (2023). *Cashless in China? A problem no more for foreign travelers.* Retrieved from https://www.travelweekly-asia.com/Travel-News/Travel-Technology/WeChat-Pay-and-Alipay-to-accept-global-credit-cards. Assessed 4 February 2024.

Un, C., Thinyane, M., & Christine, D. (2021). *Civil society organizations' cyber resilience.* United Nations University, Institute in Macau.

Van Veldhoven, Z., & Vanthienen, J. (2023). Best practices for digital transformation based on a systematic literature review. Digital Transformation and Society, (ahead-of-print) https://doi.org/10.1108/DTS-11-2022-0057

Word Bank Report. Doing Business. (2019). Training for Reform. 16th edition. A World Bank Group Flagship Report. Retrieved from https://www.ngbusiness.org/content/dam/ngBusiness/media/AnnualReports/English/DB2019-report_web-version.pdf. Assessed 20 July 2019.

World Population Review. (2024). *Countries in Asia 2014.* Retrieved from https://worldpopulationreview.com/country-rankings/countries-in-asia. Accessed 5 February 2024.

Choi Chea Chiam earned Bachelor of Economics (Honours) in 2001,Master of Business Administration (MBA) in 2003 and PhD (Economics) in 2014 respectively. Starting in early 2003, she began her lecturing career at an international private university college and currently at a private university. With approximately 20 years of academic experience, her research interests focus on environmental, resource, social and recreational economics. In addition, she is also interested in researching latest learning technologies in open and distance learning, as well as lifelong learning.

Chapter 10
Future of Shipping in Asia: China and Singapore's Coping Mechanisms in Supply Chain

Clusters of Future Studies: Corporate Foresight; Post-normality and Complexity; Technological Trends

Wan Xin Wu

Abstract In the contemporary world of twenty-first century, globalisation and trade have become crucial factors to facilitate growth of economies and cooperation between different stakeholders. Following the rise of Asia, the region has begun to prosper and progress in terms of supply chain resiliency and operations. Notably, some of the busiest ports in Asia are in China and Singapore. In this paper, a case study of both countries will be conducted to understand how coping mechanisms can be used to mitigate challenges faced by Asian countries, particularly freight and logistics management. Over the years, China has developed with incredible speed to become an emerging global economic powerhouse and holds a significant wealth of supplies. As a small state lacking natural resources, Singapore has managed to redefine its hinterlands and build a reputation of being an internationally recognised financial hub. The movement of cargoes is especially important in critical times, such as during the COVID-19 pandemic and even a series of natural disasters witnessed in 2022 like earthquakes, floods and wildfires. These circumstances have impelled observers to speculate and deliberate about how best to tackle problems that may be out of man's control. Moreover, the paper will attempt to provide some insights into Asia's shipping industry and how it promotes countries' reputation in the international arena to foster positive relations with other key actors.

Keywords Digitalisation · Green solutions · Port infrastructure · Shipping industry · Supply chain resiliency · Sustainability

W. X. Wu (✉)
Independent Researcher, Singapore
e-mail: wuwx2022@gmail.com

M. S. Balaji et al. (eds.), *Future of Business in Asia*,
https://doi.org/10.1007/978-3-031-95940-0_10

10.1 Introduction

In an increasingly volatile and fast-paced global environment, supply chain resiliency and operations are particularly of essence to facilitate movement of goods and services. The process of ensuring smooth transportation is crucial to save time, cost and even human lives. Following the rise of Asia, ports in the region have become busier than before, as more cargoes are passed through these locations. Yet, it is inevitable for unexpected problems to surface along with more opportunities. The challenges can posit themselves in different forms, be it natural or man-made. This chapter focuses on exploring China and Singapore's coping mechanisms in supply chain to provide insights in Asia's shipping industry and how they promote the countries' reputation in the international arena.

This chapter starts by highlighting how qualitative case study approach is employed, focusing on comparative analysis of shipping dynamics in China and Singapore. Following this, it examines the rise of shipping in Asia and its increasing relevance with reference to globalisation and trade. A discussion on reasons for choosing China and Singapore as case studies to understand freight and logistics management will then be presented. Subsequently, a critical analysis is provided to assess how strategies are being implemented by the two Asian countries, outlining their successes and challenges. The chapter will conclude with reflective thoughts on the coping mechanisms in Asia's shipping industry and how workflow processes may be expected to evolve over time in the long run.

10.2 Literature Review

10.2.1 Rise of Shipping in Asia: Globalisation and Trade

With the advancement of technology as well as interconnectedness, observers have noticed vast changes in the economic and social aspects. Processes in businesses are being more streamlined and integrated as cooperation between different stakeholders rises to a whole next level as compared to decades ago.

In the contemporary world of today, the global system on which the world operates is heavily influenced by the impact of communications and mechanically powered transportation, with 'economic complementarities and exchanges' as foundation (McNeill, 1990, 10). The bureaucratic systems implemented by different nations depend on established trade networks that transcended political boundaries, which can be related to ancient times of pre-existing caravans and shipping connections (McNeill, 1990, 12). Through modernisation and industrialisation, the 'rise of Asia' phenomenon (particularly observed via China's rapid growth) has emerged with the expansion of Asian economies alongside increased geopolitical significance. Notably, the 'shipping industry plays an important role in stoking economic development' (Wong et al., 2009a, 2009b, quoted in Lai et al., 2011, 632).

To trace developments of shipping in Asia, the global shipbuilding industry was seen to face rising competition from the East back in 1950s; as the Japanese entered the scene, followed by the Koreans and the Chinese subsequently in 1990s (Tenold, 2019). According to the United Nations Conference on Trade and Development (UNCTAD)'s Review of Maritime Transport published in 2021, Asia plays a predominant role in global maritime trade and remained the largest supplier of seafarers worldwide. This is a notable perception to emphasise how Asian countries are contributing to the shipping industry by facilitating supply chain processes and providing human resources.

The growth of manufacturing industries in Asian countries has resulted in an increase in exports, which eventually contributed to the aggregate demand. Identifying as a strategic geolocation for global trading hubs, Asia functions as facilitators promoting intercommunication, with shipping being an important sector to boost economic growth and cooperation. To elaborate, the Review of Maritime Transport released by UNCTAD in 2023 indicated that Asia houses economies with the highest level of connectivity in terms of linear shipping based on the Liner Shipping Connectivity Index (LSCI). In brief, according to the World Bank's DataBank, LSCI assesses the linkages between different countries based on shipping activities to highlight the maritime trade connections (The World Bank, 2023). Additionally, the Container Port Performance Index (CPPI) produced by the World Bank and S&P Global Market Intelligence has indicated that 18 out of the top 25 ports globally are located in Asia. Hence, these circumstances portray how Asia has become a central player in the shipping industry with the optimal settings for the flourishing of commerce.

10.2.2 Why Are China and Singapore Chosen as Case Studies?

Using China and Singapore as case studies, this chapter attempts to provide an understanding of how Asian economies execute and carry out freight and logistics management in daily workflow. By examining the processes and movements involved in the shipping industry, it provides an insight into how cost-effective measures are implemented to ensure supply chain resiliency and visibility in the long-run.

To briefly explain, China and Singapore are chosen in this analysis as both countries host some of the world's busiest ports and have continuously invested significant amounts of resources and efforts in upgrading and maintaining port infrastructures. According to World Shipping Council (WSC, 2023) which represents the global liner shipping industry, Port of Tianjin and Port of Singapore were ranked among the Top 10 container ports in 2021, with the volume of containerised cargo hitting 20.27 Million twenty-foot equivalent units (TEUs) and 37.49 Million TEUs respectively. Also, both ports ranked among the top 25 for container port performance in the CPPI 2023, based on a report produced by the World Bank (The World Bank, 2024).

Particularly in ocean freight, China and Singapore utilise advanced technologies to remain competitive in the shipping industry while partaking in cross-regional cooperative initiatives. By demonstrating the commitment to maritime trade, the nations may be viewed as transportation middlemen to facilitate the flow of trade (Wong et al., 2009a, 2009b; Yang et al., 2009) amid complexities in supply chain risk management.

Over the years, China has developed with incredible speed to become an emerging global economic powerhouse and holds a significant wealth of supplies. The expanding power and influence that the nation wields has contested the idea of those who perceived the status quo in the West as stable and secure (Levesque, 2011, 4). The transformation of the global maritime trade landscape with China championing the race has affected the longstanding dominance of Western nations, and made it difficult for them to sustain their customary influence and success (Schober, 2021). The Chinese view infrastructure development as the approach for attaining several strategically interconnected goals such as economic growth, enhanced competitiveness and fulfilling national aspirations (Levesque, 2011, 14). Hence, the shipping industry is crucial in broadening our understanding of the projection of China's economic outlook.

On the other hand, despite being a small state lacking natural resources, Singapore has emerged and successfully redefined its hinterlands while building a reputation as an internationally recognised financial hub. The nation serves as a reference point that size limitation does not necessarily equate to hindrance to economic growth. The island has upheld the essence of being a port city, which evolved and modernised via leveraging technology and embracing innovations strategically to sustain its position as one of the world's busiest ports (Tan, 2006, 857). Thus, analysing how the country remains relevant in the shipping industry is useful in identifying suggestions for mitigating possible challenges.

The shipping industry may well serve as a reflection to deepening our understanding of the core of business practices between two countries that have varying geographical sizes and power. While China focuses on handling diverse bulk goods via digitalisation as a strong key player, Singapore stresses on connectivity and green solutions. With technology as the basis, both countries attract investors' confidence with different motivations.

10.3 Methodology

This chapter utilises a qualitative case study approach, focusing on comparing the different shipping dynamics in China and Singapore. While primary data collection is not involved, it uses secondary data to highlight coping mechanisms employed by both countries in their supply chains to ensure resiliency and risk management. This will encourage an in-depth understanding of publicly available data from various sources as well as insights in operational strategies of the two significant players in the shipping sector. Some criteria that were taken into consideration include credibility of the sources, publication dates, as well as how nuanced the perspectives are presented in the data sets. Meanwhile, this study acknowledges certain limitations in using secondary data, such as potential biases and how there might be gaps in data availability regarding shipping operations that might not be released publicly.

10.4 Findings

10.4.1 Case Study of China: Leading Advanced Port Solutions in Dynamic Global Environment

Being home to at least 34 major ports and more than 2000 minor ports (Wang & Wang, 2019), China is an important case study to understand how the country managed to find success in this profitable industry and build a name for itself in the international arena. Since the list of ports is rather exhaustive, this chapter attempts to highlight a crucial port (namely Port of Tianjin) of the nation to understand how we can learn from it as it remains relevant in the ever-changing world of today.

In the current phase of digitalisation, there is heavy emphasis on advanced technologies, automation and smart systems. Similar to other sectors, the maritime industry undergoes digital transformation involving intelligent and autonomous shipping solutions (Johns, 2018). As we witnessed the rise of Industry 4.0 (Durán et al., 2019), the implementation of a smart port, which is intentionally designed and equipped with technologies to upgrade port performance, can tackle emerging problems encountered by global trade and logistics networks (Li et al., 2023, 1).

To delve deeper, the Port of Tianjin is an example to illustrate how labor-intensive work has been cut down and replaced by technologies on a great scale. The port is collaborating with Huawei Technologies and other partners to address challenges within the port industry, which includes enhancing resilience of supply chains via leveraging expertise in 'communication technologies, artificial intelligence and autonomous driving' (The Straits Times, 2023). Furthermore, at the 2023 Smart Port Conference, Huawei Technologies has announced to work with Tianjin Port (Group) Company, Zhejiang Seaport Group, National Supercomputing Center of Tianjin and CloudWalk Technology Company to jointly initiate research and development of the large-scale model of PortGPT (Li, 2023). This will aid in improving service capabilities for port enterprises, especially with production shift meetings and other general issues at Port of Tianjin (Li, 2023). Hence, we are able to witness the coupling of merging technological means and problem-solving skills to strengthen workflow processes effectively while supporting decision-making in port operations. This is particularly critical in pandemic times and when encountering natural disasters like hurricanes and floods.

10.4.2 Case Study of Singapore: Connoisseur of Communication and Advocate of Green Solutions

Being a small state with finite resources, Singapore has been proactively seeking ways to redefine its hinterlands while navigating to survive. For definition purpose, small states are 'entities which are exceptionally small in area, population and human and economic resources' (UNITAR, 1971, 29). Despite the limitations in terrestrial landscape, Singapore plays a crucial role as port of call for many maritime trading routes spanning both East–West and North–South axes, as well as a central point for cargoes heading to neighbouring ports (Tran et al., 2022, 306). Being interconnected with more than 600 other ports (PSA Singapore, 2023), Singapore has extensive established links with other stakeholders that provides the nation-state with flexibility and competitive advantage in the global marketplace.

Following the post-colonial era, the country has undergone remarkable economic transformation to become a successful commerce centre due to strategic investment in cutting-edge technologies and progress in advanced industrial production (Mindur, 2020, 109). With the shift towards using larger container ships in an attempt to meet the rising demand of ocean freight, Singapore has positioned itself as one which can cater to the needs of large vessels due to thorough strategising of infrastructure development (Mindur, 2020, 116). Furthermore, the Port of Singapore also provides services for ships passing through the Strait of Malacca with regards to maintenance and supply resources (Mindur, 2020, 118), which largely entices others to make a stop at the port. This encourages potential clients to also consider Singapore as a destination for loading and unloading goods, and refueling among other maritime-related activities.

10.5 Discussion of Implications

10.5.1 Impact of COVID-19 and Natural Disasters on Port Operations in China

Looking back when COVID-19 struck in 2019, shipping networks have experienced significant disruptions affecting the global supply chain system (Jin et al., 2022, 90). When China introduced its iron-fist 'Zero COVID' containment policy to prevent spread of the coronavirus, lockdowns and mass testing were implemented. This affected port operations tremendously, particularly in terms of labor supply and led to higher risks of backlog issues arising from local authorities' restrictions (Xie, Yang, & Strumpf, 2022). The situation has underscored the vulnerability and deficiency of the container market, which lacked the will to adapt and respond to crises in times of emergency (Chen et al., 2022).

When faced with natural disasters such as typhoons and flooding, operations at the port can be impacted on a huge scale. Disturbances at the port resulting from these natural phenomena can have regional and national consequences (Adam et al., 2016). Earlier on in August, lingering effects of Typhoon Doksuri brought upon severe flooding, causing rivers in Tianjin to exceed warning levels and around 35,000 people were evacuated from near the affected Yongding River (Al Jazeera, 2023). Experts had commented that such occurrences in Northern China had disrupted operations at Port of Tianjin and anticipated recovery efforts to take up to an additional two to four weeks (Margaronis, 2023). When the wind force in Port of Tianjin reached level 7, governing authorities had come up with temporary traffic control measures to allow one-way navigation to ensure safety of ship navigation, as well as initiating emergency plans to utilise the comprehensive maritime digital management system (Zhang, 2023). To add on, earthquakes may be triggered by tropical cyclones, and these can result in even greater damages to operations at the port.

To briefly supplement, though wildfires may not impact port operations in the most direct manner, the occurrences can lead to reduced visibility in the port when smoke particles spread; road closures in surrounding areas which affect transportation of goods; and damaging other nearby critical power infrastructures.

10.5.2 Singapore's Response to Sustainability and Crisis Management

Singapore has attempted to support the net zero commitments on reducing greenhouse gas emissions to combat climate change. The Port of Singapore Authority (PSA) will implement a 'smart grid management system and construct green buildings', of which an example is that the Tuas Maintenance Base Administrative Building will use 58 percent less energy while producing solar energy to offset electricity usage (Maritime and Port Authority of Singapore, 2023). This aligns with what the United Nations has called upon for global efforts among countries to play their part in the global effort for preservation of a 'livable planet' (United Nations, 2023), which stems from the 2015 Paris Agreement. As such, Singapore has shown itself to be an active global citizen in international cooperative initiatives. This is particularly important for small states like Singapore that strongly advocate respecting sovereignty of states as well as an active contributor to clauses raised by international organisations. Hence, how Singapore runs its port can be viewed as an extension of soft power demonstration to achieve a reputable status in the shipping industry. Soft power diplomacy serves as a useful policy instrument for states to leverage their assets for collaboration with others in order to drive mutual benefits and foster meaningful state-relations.

Revisiting the challenges that Port of Singapore had experienced during the COVID-19 period, officials have observed that there was a reduction in the container throughput by 0.9 percent for 2020 amid overseas port closures, shipping delays and worldwide shortage of containers (Kok, 2021). Yet, Singapore has continued to expand in its maritime sector in the same year, with 11 new shipping groups establishing their presence in the country and eight existing ones broadening business activities (Kok, 2021). Governing authorities have mentioned that such reason may be due to the port's introduction of contactless cargo and bunkering procedures while offering telemedicine services for seafarers (Kok, 2021).

10.6 Future Directions

10.6.1 Building Port Resilience in China Through Technology and Strategic Policy Measures

With robots working at the driverless smart terminal co-built by Huawei as well as utilising only 25 percent of human resources to manage neighbouring docks (The Straits Times, 2023), the challenges faced by ports due to lack of manpower may be resolved and damages even minimised. This will lower the risk of port disruption that can affect the supply chain, especially with container terminals that handle large volumes of cargoes. Additionally, it can mitigate problems of processing delays in situations when governing authorities impose more stringent curbs.

Countries that rely heavily on workers to operate the ports can consider such business models that invest in technology to reduce pressure on the need to ensure constant supply of personnel-on-ground while depending on automation for greater efficiency. This also further promotes remote working as human resources can be redistributed to work in other areas that require higher skill sets and expertise while providing chances of upgrading to retain existing talents.

Meanwhile, on an applaudable move by the government in an attempt to overcome the impact of the pandemic and restore operations at Port of Tianjin, we can refer to the notice issued by the Tianjin Municipal Prevention and Control Headquarters in March 2020. One notable point mentioned was that containers transshipped by foreign trade container shipping companies at Port of Tianjin will be exempted from terminal yard usage fees from March 17 to June 30, 2020 (Tianjin Municipal Prevention and Control Headquarters, 2020). This suggested that authorities had attempted to aid not just domestic but also foreign enterprises to demonstrate the Chinese's goodwill. It also paves way for future economic collaboration post COVID-19 and boosts China's reputation as one which promotes cooperation and empathy while emerging as a strong key player in the maritime sector.

Thus, it is crucial for port authorities to be aware and prepare for dire circumstances by coming up with thorough backup plans taking into consideration safety hazards as well as minimising possible casualties and even announcing force majeure events when necessary to carry out effective communication to respective stakeholders.

10.6.2 Enhancing Innovation and Collaboration in Singapore's Maritime Future

Notably, Singapore's Tuas Port which officially opened in September 2022, is projected to handle up to 65 million TEUs upon targeted completion in the 2040s (Maritime and Port Authority of Singapore, 2023). Like China's Port of Tianjin in terms of utilising modern technology, Tuas Port will deploy artificial intelligence (AI) and automation processes in operations. There is the ongoing development of Next Generation Vessel Traffic Management System to offer precise and real-time awareness of the traffic situation, which can streamline processes and optimise turnaround time of ships for smoother facilitation (Maritime and Port Authority of Singapore, 2023). This serves as a strategy of risk management to practise predictive monitoring coupled with advanced analytics. Additionally, the port will adopt 'versatile and maneuverable electrified automated yard cranes and automated guided vehicles (AGVs)' as well as remote management from Tuas Port Control Centre to enhance overall work productivity and coordination (Maritime and Port Authority of Singapore, 2023).

On promoting innovations to resolve maritime issues, the Smart Port Challenge organised by Port Innovation Ecosystem Reimagined @ Block71 attempts to highlight problem statements and seeks out technology start-ups that can receive funding to lead the move and provide effective maritime solutions (Kok, 2021). Additionally, authorities have indicated that Maritime and Port Authority of Singapore (MPA) are committed to collaborating with employers, the Singapore Shipping Association, and trade unions to roll out new company training committee initiatives that facilitate digitalisation and training plans for workers (Kok, 2021). Such ventures allow the country to 'think outside the box' and be more open to plausible ideas from talents with diverse knowledge and backgrounds. This also prompts professionals to be more motivated in contributing to the maritime sector while receiving affirmation that the country values contributions in this knowledge-based economy.

For cooperative measures, in December 2023, Singapore signed a Security of Supply Arrangement with the United States, for which the agreement seeks to aid the countries in managing unexpected supply chain disruptions (Singh, 2023). Under the arrangement, the two nations can 'request priority delivery of critical and urgent industrial resources to meet their defence needs' (Singh, 2023). This is noteworthy and certainly one way to show how Singapore is 'preparing for the rainy days'. As a backup plan, this will provide an alternative for the country to mitigate possible upcoming challenges in not just port operations but also other aspects of logistical issues. Without effective diplomacy, this will be a difficult move since skillful communication is essential for discussion to proceed in time. Taking into account a combination of defence and business aspects, and translating the outcome into a beneficial one requires efforts on both sides to plan strategically while understanding the pros and cons. In a long-term consideration, this will make Singapore a more attractive partner for collaboration and draw more investments for new opportunities.

Thus, the above points highlight how Singapore's potential to thrive as an important port is not restricted due to size, but rather it is able to build a reputation in the international arena through effective management and cooperation with other stakeholders while driving green solutions. As the 'middleman' in the logistical process, the country is aware that choke points at other connected seaports could lead to disruptions of deliveries and shipping routes. By investing strategically in apt and sustainable domestic shipping practices, Singapore can at least reduce risks while doing scenario planning since external factors tend to be volatile and even unprecedented at times.

10.7 Conclusion

All in all, Asia's shipping industry is booming with opportunities and challenges at the same time. With the region becoming an even more attractive spot for manufacturing and other economic activities, it is important for countries to identify their strengths and existing problems. In a world where change is constant, governing authorities, businesses and other stakeholders should collaborate and come up with backup plans in times of emergencies. For scenarios that are out of man's control, it is crucial to identify the possible risks involved and attempt to utilise strategies that minimise damage to affecting parties. As the world advances, the implementation of technologies and artificial intelligence in the shipping industry will become a norm in time to come. The growing need for such solutions is expected to persist (Kok, 2021) and it is recommended to invest in building up these applicable skills to continue staying competitive. However, these cannot be done without proper budgeting and competent administration of the states to ensure that a portion of funds is allocated to the respective sector. Likewise, businesses have important roles to play in practicing corporate social responsibility to contribute to the country's growth. Additionally, maintaining positive state-relations with other nations is also necessary for technology transfers, and learning from those who are already ahead in the 'shipping race'. Nevertheless, there is no one-size-fits-all solution and stakeholders should practice flexibility while extending their foresight to encourage innovation and adaptation in today's volatile global society.

Highlights

- China and Singapore lead Asia's business future with artificial intelligence, automation, and digital innovation in port management
- China enhances supply chain resilience through strategic policies and cutting-edge technology

- Singapore spearheads sustainable maritime practices, shaping Asia's green shipping future
- Both nations foster a tech-driven, collaborative, and sustainable business environment in Asia.

References

Adam, E. F., Brown, S., Nicholls, R. J., & Tsimplis, M. (2016). A systematic assessment of maritime disruptions affecting UK ports, coastal areas and surrounding seas from 1950 to 2014. *Natural Hazards, 83*, 691–713. https://doi.org/10.1007/s11069-016-2347-4

Al Jazeera Media Network. (2023, August 4). China floods threaten downstream cities as tens of thousands evacuated. *Al Jazeera Media Network*. https://www.aljazeera.com/news/2023/8/4/china-floods-threaten-downstream-cities-as-tens-of-thousands-evacuated

Chen, C., Feng, T., Gu, X., & Yao, B. (2022). Investigating the effectiveness of COVID-19 pandemic countermeasures on the use of public transport: A case study of The Netherlands. *Transport Policy, 117*, 98–107.https://doi.org/10.1016/j.tranpol.2022.01.005

Durán, C. A., Córdova, F. M., & Palominos, F. (2019). A conceptual model for a cyber-social-technological-cognitive smart medium-size port. *Procedia Computer Science, 162*, 94–101. https://doi.org/10.1016/j.procs.2019.11.263

Jin, L., Chen, J., Chen, Z., Sun, X., & Yu, B. (2022). Impact of COVID-19 on China's international liner shipping network based on AIS data. *Transport Policy, 121*, 90–99. https://doi.org/10.1016/j.tranpol.2022.04.006

Johns, M. (2018, October). Seafarers and digital disruption—The effect of autonomous ships on the work at sea, the role of seafarers and the shipping industry. Retrieved December 27, 2023 from https://www.ics-shipping.org/wp-content/uploads/2020/08/ics-study-on-seafarers-and-digital-disruption-min.pdf

Kok, Y. (2021, June 18). Port of Singapore regains momentum despite disruptions caused by Covid-19 pandemic. *The Straits Times*. https://www.straitstimes.com/singapore/singapore-port-handled-more-cargo-this-year-despite-disruptions-caused-by-covid-19

Lai, K., Lun, V. Y. H., Wong, C. W. Y., & Cheng, T. C. E. (2011). Green shipping practices in the shipping industry: Conceptualization, adoption, and implications. *Resources, Conservation and Recycling, 55*(6), 631–638. https://doi.org/10.1016/j.resconrec.2010.12.004

Levesque, P. J. (2011). *The shipping point: The rise of China and the future of retail supply chain management*. John Wiley & Sons

Li, D. (2023, October 19). 华为联合天津港集团等多家单位联合启动港口大模型PortGPT研发 [Huawei, in collaboration with Tianjin Port Group and multiple other units, jointly initiates the development of the large-scale PortGPT model]. *The Securities Times*. https://www.stcn.com/article/detail/1009139.html

Li, K. X., Li, M., Zhu, Y., Yuen, K. F., Tong, H., & Zhou, H. (2023). Smart port: A bibliometric review and future research directions. *Transportation Research Part e: Logistics and Transportation Review, 174*, 1–15. https://doi.org/10.1016/j.tre.2023.103098

Margaronis, S. (2023, August 17). Port of Tianjin recovery from storms & floods may take 2 to 4 weeks. *American Journal of Transportation (AJOT)*. https://www.ajot.com/insights/full/ai-port-of-tianjin-recovery-from-storms-floods-may-take-2-to-4-weeks

McNeill, W. H. (1990). "The Rise of the West" after Twenty-Five Years. *Journal of World History, 1*(1), 1–21. http://www.jstor.org/stable/20078455

Mindur, M. (2020). Significance of the port of Singapore against the country's economic growth. *Scientific Journal of Silesian University of Technology, Series Transport, 106*, 107–121. https://doi.org/10.20858/sjsutst.2020.106.9

MPA (Maritime and Port Authority of Singapore. (2023). *Tuas Port—A smarter and greener port*. Port of Singapore. Retrieved December 27, 2023, from https://www.mpa.gov.sg/maritime-singapore/port-of-the-future

PSA (Port of Singapore). (2023). *Numbers at a Glance*. Port of Singapore. Retrieved December 29, 2023, from https://www.singaporepsa.com

Schober, E. (2021). Building ships while breaking apart: Container economies and the limits of chaebol capitalism. *Focaal, 2021*(89), 12–24. https://doi.org/10.3167/fcl.2021.890102

Singh, K. (2023, December 7). Singapore, US sign defence supply chain agreement. *The Straits Times*. https://www.straitstimes.com/singapore/singapore-us-sign-defence-supply-chain-agreement

Tan, T. Y. (2006). Port cities and hinterlands: A comparative study of Singapore and Calcutta. *Political Geography, 26*(7), 851–865. https://doi.org/10.1016/j.polgeo.2007.06.008

Tenold, S. (2019). The declining role of western Europe in shipping and shipbuilding, 1900–2000. In N. P. Petersson, S. Tenold, & N. J. White (Eds.), *Shipping and globalization in the post-war era* (pp. 9–33). Palgrave Macmillan.

The Straits Times. (2023, January 17). Major China port banks on automation to beat Covid-19 disruptions. *The Straits Times*. https://www.straitstimes.com/asia/east-asia/major-china-port-banks-on-automation-to-beat-covid-19-disruptions

The World Bank. (2023). Metadata Glossary. *The World Bank Group*. Retrieved December 27, 2023, from https://databank.worldbank.org/metadataglossary/world-development-indicators/series/IS.SHP.GCNW.XQ

The World Bank. (2024). *The container port performance index 2023: A comparable assessment of performance based on vessel time in port*. The World Bank Group. Retrieved December 20, 2024, from https://openknowledge.worldbank.org/server/api/core/bitstreams/6cebb847-6f46-44e7-9533-12ac893b3693/content

Tianjin Municipal Prevention and Control Headquarters, The People's Government of Beijing Municipality. (2020, March 19). 关于天津港口支持企业克服疫情影响实现复工复产若干措施的通知 [Notice on Measures Regarding Tianjin Port's Support for Enterprises to Overcome the Impact of the Epidemic and Achieve Resumption of Work and Production]. https://www.beijing.gov.cn/zhengce/zhengcefagui/202005/t20200506_1890372.html

Tran, N. K., Lam, J. S. L., Jia, H., & Adland, R. (2022). Emissions from container vessels in the port of Singapore. *Maritime Policy & Management, 49*(3), 306–322. https://www.tandfonline.com/doi/full/https://doi.org/10.1080/03088839.2021.1980236

UN (United Nations). (2023). For a livable climate: Net-zero commitments must be backed by credible action. *United Nations*. Retrieved December 27, 2023, from https://www.un.org/en/climatechange/net-zero-coalition

UNCTAD (United Nations Conference on Trade & Development). (2021, November 2018). Asia expands its lead in maritime trade and business. *UNCTAD*. https://unctad.org/news/asia-expands-its-lead-maritime-trade-and-business#:~:text=Asia%27s%20predominance%20in%20global%20maritime,2021%20published%20on%2018%20November

UNCTAD (United Nations Conference on Trade & Development). (2023, September 27). Review of Maritime Transport 2023: Facts and Figures on Asia. *UNCTAD*. https://unctad.org/press-material/review-maritime-transport-2023-facts-and-figures-asia

UNITAR (United Nations Institute for Training and Research). (1971). *Small states and territories: status and problems. A UNITAR study by Rapaport, J., E. Muteba, and J. J. Therattil. Contributions from The Division of Public Administration, United Nations and C. L. Taylor, Yale University*. Arno Press

Wang, Y., & Wang, N. (2019). The role of the port industry in China's national economy: An input–output analysis. *Transport Policy, 78*, 1–7. https://doi.org/10.1016/j.tranpol.2019.03.007

Wong, C. W. Y., Lai, K., & Ngai, E. W. T. (2009a). The role of supplier operational adaptation on the performance of IT-enabled transport logistics under environmental uncertainty. *International Journal of Production Economics, 122*(1), 47–55. https://doi.org/10.1016/j.ijpe.2008.12.023

Wong, C. W. Y., Lai, K., & Cheng, T. C. E. (2009b). Complementarities and alignment of information systems management and supply chain management. *International Journal of Shipping and Transport Logistics, 1*(2), 156–171. https://doi.org/10.1504/IJSTL.2009.024493

World Shipping Council. (2023). The top 50 container ports. World shipping council. Retrieved December 27, 2023, from https://www.worldshipping.org/top-50-ports

Xie, S. Y., Yang, J.& Strumpf, D. (2022, January 11). China Covid-19 lockdowns hit factories, ports in latest knock to supply chains. *The Wall Street Journal*. https://www.wsj.com/articles/china-covid-19-lockdowns-hit-factories-ports-in-latest-knock-to-supply-chains-11641918247

Yang, C., Marlow, P. B., & Lu, C. (2009). Assessing resources, logistics service capabilities, innovation capabilities and the performance of container shipping services in Taiwan. *International Journal of Production Economics, 122*(1), 4–20. https://doi.org/10.1016/j.ijpe.2009.03.016

Zhang, L. (2023, August 6). 天津港主航道单向通航 [One-way navigation of main channel of Tianjin Port]. Tianjin Daily. http://epaper.tianjinwe.com/tjrb/html/2023-08/06/content_152_8109961.htm

Wan Xin Wu was a Senior Analyst with the China Programme at S. Rajaratnam School of International Studies (RSIS), Nanyang Technological University, Singapore. Prior to that, she had worked in both the corporate risk management and education sectors. She obtained her Master of Science in Asian Studies from RSIS. Her research interests include global supply chain risks, sustainability transitions, and the role of small and middle powers in shaping international norms. She is also keen to explore the intersections of international law, technology and geopolitics.

Chapter 11
Future-Proofing Business: The Changing Landscape of Internal Audit

Clusters of Future Studies: Corporate Foresight; Environment Futures

Suhaily Shahimi, Mohd Haniff Zainuldin, and Elaine Yen Nee Oon

Abstract Internal audit plays an increasingly critical role in ensuring business success, as its focus has shifted from compliance to a more strategic and advisory function (Lenz and Jeppesen, EDPACS, 2022). With the growing complexity of modern organisations, internal auditors are expected to leverage technology, data analytics, and a proactive approach to help manage emerging risks such as environmental, social, and governance (ESG) issues, enhance processes, and support strategic objectives. The ABC model, proposed by (Lenz and Hoos, Build Consult EDPACS, 2023), advocates that internal auditors should expand their traditional roles of 'Assure' and 'Consult' to also include a 'Build' function. By doing so, auditors can actively engage in organisational projects, contributing to the architectural aspects of business operations and addressing the complexities of today's environment. This expanded role allows internal auditors to create value by aligning their efforts with crucial areas like ESG strategy, IT governance, and risk mitigation. The model emphasises that integrating the 'Assure,' 'Build', and 'Consult' functions will enable auditors to provide a more comprehensive and forward-looking approach to ESG and other strategic challenges. This chapter aims to substantiate the ABC model's relevance for enhancing auditors' involvement in ESG management, offering a robust framework for addressing ESG goals and risks effectively.

Keywords Internal audit · ABC model · ESG · Future-proofing business · Risk management

S. Shahimi (✉) · M. H. Zainuldin · E. Y. N. Oon
Department of Accounting, Faculty of Business and Economics, Universiti Malaya, Kuala Lumpur, Malaysia
e-mail: suhaily@um.edu.my

M. H. Zainuldin
e-mail: haniff.zainuldin@um.edu.my

E. Y. N. Oon
e-mail: oonelaine@um.edu.my

M. S. Balaji et al. (eds.), *Future of Business in Asia*,
https://doi.org/10.1007/978-3-031-95940-0_11

11.1 Introduction

Internal audit has undergone substantial evolution throughout history, with its origins dating back to ancient civilisations where rudimentary systems of checks and balances were established to protect assets. However, the formalisation of internal audit gained momentum during the industrial revolution, spurred by the necessity for structured oversight within burgeoning organisations. The establishment of the Institute of Internal Auditors (IIA) in 1941 marked a pivotal moment, facilitating the global standardisation of internal audit practices (Gupta & Ray, 1992). In the ensuing decades, the business landscape has undergone rapid and profound transformations driven by technological advancements, globalisation, and changing regulatory landscapes, necessitating continuous adaptation within organisations worldwide.

Today, internal audit functions serve as strategic partners to organisations, providing assurance on risk management, governance, and control processes. With advancements in technology, internal auditors increasingly leverage data analytics and automation to enhance efficiency and effectiveness. Additionally, there's a growing emphasis on proactive risk identification and value-added insights to support decision-making at all levels of the organisation. Hence, it is essential for internal auditors to expand their responsibilities to include a thorough understanding and evaluation of Environmental, Social, and Governance (ESG) issues within organisational contexts (Bonrath et al., 2022). This entails integrating ESG considerations into risk assessment frameworks, evaluating the effectiveness of ESG policies and practices, and providing assurance on ESG disclosures. But to tackle intricate ESG risks, Lenz and Hoos (2023) advocate for integrating a 'building' role into internal audit. This empowers auditors to participate in creation processes, fostering company resilience and growth. When combined with 'assurance' and 'consulting', this model enables auditors to guide firms through ESG challenges towards innovation. Additionally, internal auditors should enhance collaboration with sustainability and corporate social responsibility (CSR) functions, leverage data analytics to identify ESG-related risks, and provide insights to support strategic decision-making. Embracing ESG complexities aligns internal audit with broader organisational objectives, enhances stakeholder trust, and contributes to sustainable value creation (IIA, 2021).

In Asia, the internal audit landscape presents unique challenges and opportunities owing to the region's diverse regulatory environments, economic development stages, and cultural dynamics. Many Asian companies, particularly those in emerging economies, are facing rapid technological changes, evolving ESG expectations, and increasingly complex supply chains. Therefore, internal auditors in Asia must adapt by integrating proactive strategies such as the ABC model to remain relevant and effective. This chapter is particularly significant because it addresses these regional complexities, offering a framework that supports Asian auditors in navigating ESG integration, digital transformation, and stakeholder engagement within their respective markets.

The first section of our chapter begins with the background of the internal audit function from the historical context until the present. In the next section (Sect. 11.2), we describe the current internal audit landscape, providing an overview of past studies regarding the internal audit role and function; and justifying the need for a strategic model, namely the Assure-Build-Consult (ABC) model. This is followed by the definition and elaboration of the ABC model. In Sect. 11.3, we explain how internal auditors utilise the proposed ABC model in executing their internal audit roles and functions. Next, we provide examples of how internal auditors from organisations in three different industries (oil and gas; financial; and retail) utilise the ABC model in their ESG-related risk assessment, reporting requirements, and IT and supply chain governance. In the final section (Sect. 11.4), we conclude by illustrating the contributions of this chapter, limitations, and future directions of integrating the ABC model within the internal audit role and function.

11.2 Key Concepts and Issues

11.2.1 Evolution of Internal Audit: Current Landscape and Strategic Model

11.2.1.1 *The Current Internal Audit Landscape and the Need for a Strategic Model (ABC Model)*

According to the Institute of Internal Auditors (IIA), the role of internal audit is 'to provide independent assurance that an organisation's risk management, governance and internal control processes are operating effectively'. In other words, the process of internal auditing 'is an independent, objective assurance and consulting activity designed to add value and improve an organisation's operations. It helps an organisation accomplish its objectives by bringing a systematic, disciplined approach to evaluate and improve the effectiveness of risk management, control, and governance processes' (IIPF, 2017).

A recent review of the academic literature on internal audit reveals that opinions differ about the expected or actual roles of internal audit, hence no consensus was found regarding the roles internal audit should play or even its actual role in organisations (Kotb et al., 2020; Roussy & Perron, 2018). Although most studies consider the internal audit function as an oversight governance mechanism (Gramling et al., 2004; Prawitt et al., 2009; Shuwaili et al., 2023; Stewart & Subramaniam, 2010), other studies argue that the roles of internal audit are complex and constantly evolving (Handoyo, 2024; Rakipi & D'Onza, 2023; Soh & Martinov-Bennie, 2011) and therefore, cannot be confined to governance oversight. For example, Lenz and Sarens (2012) contend that there is no single, obvious solution regarding the role of internal audit and how it adds value. They argue that the uncertainty surrounding its role stems from the fact that internal auditors serve a variety of customers, including managers, the board, and the audit committee. Recognising the issue, Roussy (2013) discovered that internal auditors serve two important roles, as 'protectors' and 'helpers' of the organisation and its senior management. Meanwhile, according to Sarens et al. (2009), the internal audit function is a significant source of 'comfort' for audit committee members, whereas Vinnari and Skaerbaek (2014) emphasise that the internal audit role is central to risk management. Taken together, the literature seems to infer that the internal audit function has become the 'jack of all trades, master of none' of governance. Such an impression may create confusion about the nature of internal audit and may impair the internal audit function's perceived (or actual) added value.

In an era marked by rapid technological advancements, global interconnectedness, and dynamic market conditions, businesses today face unprecedented challenges that demand innovative approaches to risk management, control and governance. Internal audit, traditionally perceived as a compliance-oriented function and a governance mechanism that mitigates the agency problem for shareholders' benefit (Gramling et al., 2004; Prawitt et al., 2009), needs to adapt to the evolving landscape of business in order to remain effective and relevant. Additionally, a number of studies have shown that the internal audit function is a crucial organisational actor in key strategic activities such as organisational learning, training for aspiring senior managers, IT governance, and sustainability (Amoako et al., 2023; Christ et al., 2021; Roussy & Perron, 2018; Shuwaili et al., 2023). Even though sustainability concerns are projected to increase in the upcoming years (Kotb et al., 2020; Ridley et al., 2011; Soh & Martinov-Bennie, 2015), very few studies advocate that internal audit should provide assurance on sustainability (DeSimone et al., 2021). Its issues are still not well-defined and are regarded as less significant than other governance problems (Ridley et al., 2011; Soh & Martinov-Bennie, 2015). Moreover, internal auditors are perceived to have considerably less experience in environmental issues than in other areas (Amoako et al., 2023; Soh & Martinov-Bennie, 2015). Consistent with this view, a recent article by Lenz and Hoos (2023) lamented that 'like in the world of animals, the internal audit function is in a state of freeze response when it comes to ESG topics, and only slowly takes (insufficient) actions. The ESG challenge is so big, and the threats for the role of internal audit function are so real, that the profession reacts like many animals in the face of a threat: they freeze.'

Consequently, rethinking the internal audit role and function is needed to fully exploit the value proposition of the internal auditing profession. To future-proof business operations, the current internal audit role requires a strategic and proactive approach to risk management, leveraging technology, collaborating with other functions, while at the same time, ensuring compliance with regulations. Thus, internal auditors should be viewed as enablers of learning and change as well as 'builders', as proposed by the Assure-Build-Consult (ABC) model by Lenz and colleagues (Lenz & Chesshire, 2023; Lenz & Hoos, 2023).

11.2.1.2 What is the ABC Model?

The ABC model (Assure-Build-Consult) represents a paradigm shift in the traditional role of internal audit, offering a holistic framework that extends beyond mere assurance functions. The integration of 'Build' into the 'Assure' and 'Consult' aspects reflects the dynamic nature of modern organisations and the need for internal auditors to play a proactive role in contributing to organisational success and future-proofing business.

The essence of internal auditing involves assessing an organisation's governance, internal controls, and risk management procedures through objective assurance and consulting activities, aiming to ensure both legal compliance and operational efficiency (IIPF, 2017). The two main components, which are assurance and consulting play pivotal roles in ensuring the effectiveness of internal auditing. Assurance services mainly focus on providing senior management and the board with an unbiased evaluation of the adequacy and performance of internal controls. Conversely, consulting services are discretionary activities designed to enhance an organisation's operations by offering customised advice and assistance to meet its specific objectives (Betti et al., 2023; DeSimone et al., 2021). However, in addressing the intricate and unpredictable risks and challenges associated with ESG factors, Lenz and Hoos (2023) propose the incorporation of an additional function within the internal audit framework, termed 'Building'. This entails empowering internal auditors to actively engage in the processes of creation, co-creation, and construction, thereby contributing to the advancement of organisational growth and resilience (Lenz & Hoos, 2023). By leveraging the integrated model encompassing 'assurance', 'building', and 'consulting', internal auditors have the capacity to enable organisations to adeptly navigate the journey pertaining to ESG considerations and unlock new opportunities for growth and innovation.

Assure

The assurance role within internal audit is essential for evaluating the effectiveness of an organisation's governance, risk management, and control processes. It aims to identify operational inefficiencies, financial discrepancies, and non-compliance issues, thereby supporting the achievement of strategic objectives (DeSimone et al., 2021; Eulerich & Lenz, 2020). Through assessing internal controls, verifying financial report accuracy, and ensuring legal compliance, assurance activities provide recommendations for enhancements. This role is vital for promoting transparency, accountability, and continuous improvement, ultimately fostering stakeholder confidence and supporting strategic decision-making. However, the complexity and uncertainty of the ESG journey underscore the importance of assurance in validating the reliability of ESG reporting and ensuring that ESG practices align with organisational values and objectives (Lenz & Jeppesen, 2022). Constant dynamic changes make it difficult for internal auditors to anticipate new risks and adjust their audit plans in real time. Inaccurate or outdated data can hinder the relevance and accuracy of audit findings.

Build

Lenz and Hoos (2023) recommend incorporating 'building' into the scope of internal auditors' work to enhance the current assurance and consulting roles. The building function within the internal auditing process is to ensure organisations not only comply with relevant laws and regulations but also operate efficiently and effectively in achieving their strategic objectives. This function focuses on constructing robust frameworks that safeguard assets, enhance the reliability of financial reporting, and ensure compliance with applicable laws and standards. Internal auditors, through their building function, actively contribute to the development and enhancement of internal controls and processes by identifying weaknesses and recommending improvements. This proactive approach helps organisations to mitigate risks, optimise performance, and adapt to changing environments, thereby building a foundation for sustainable growth and success.

To address the intricate ESG challenges, it is necessary for internal auditors to expand their roles beyond just providing assurance and consulting services. For instance, they can lead in crafting ESG programs. Their contributions as co-creators and builders within the ESG framework make them valued partners in this process. Their education and commitment to continuous improvement position them as catalysts for learning and change within organisations, making them a valuable resource for ESG implementation (Lenz & Chesshire, 2023; Lenz & Hoos, 2023).

Consult

The consulting aspect of internal auditing is a dynamic and increasingly vital function that goes beyond the scope of traditional audit tasks, contributing to the enhancement and value creation within an organisation. Defined by its advisory capacity, this role enables internal auditors to utilise their expertise and knowledge to assist management in making informed decisions, improving processes, and achieving strategic goals (Alqudah et al., 2023; Betti et al., 2023). Differing from assurance services that concentrate on assessing compliance, risks, and controls, the consulting function is more cooperative, focusing on uncovering areas for enhancement and pioneering solutions to intricate issues. By leveraging their in-depth understanding of the organisation's operations, internal auditors in consulting roles team up with management to create plans that make the organisation run smoother, handle risks better, and become stronger and more adaptable.

However, the dynamic landscape of ESG necessitates a continuous learning process for auditors, to not only grasp the current state of risks but also anticipate future shifts that could impact organisational objectives and sustainability goals. This vigilance is crucial for internal auditors to effectively guide and advise organisations in developing robust strategies for managing ESG risks.

Figure 11.1 depicts the ABC model in internal audit. This figure illustrates the ABC Model, which expands the traditional internal audit function beyond assurance to include building and consulting roles. The model highlights how internal auditors

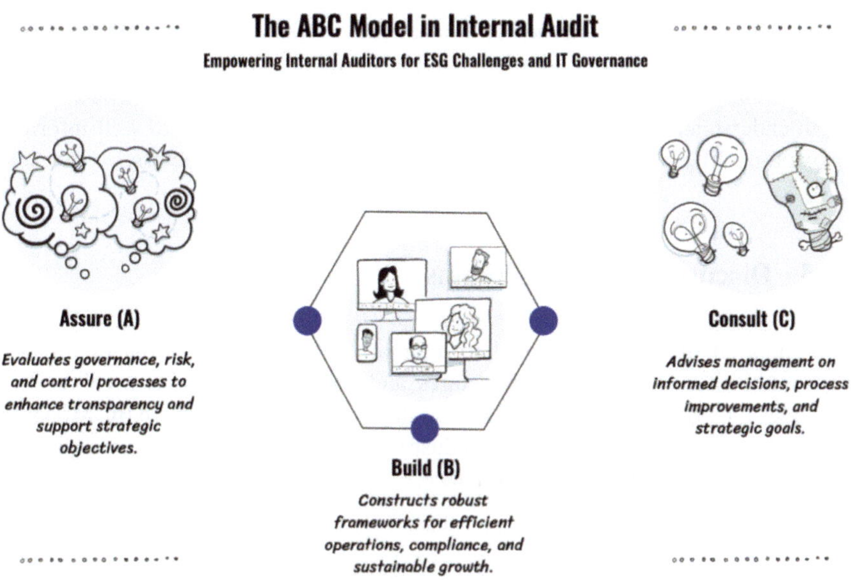

Fig. 11.1 The ABC model (Assure-Build-Consult) in internal audit

contribute to risk management, governance, and ESG strategy by integrating assurance, proactive development, and advisory functions to support business resilience and value creation.

11.2.2 Leveraging the Integrated ABC Model in Internal Audit to Future-Proof Business

The integrated framework of assurance, building and consulting (ABC model) within internal audit can be instrumental in tackling the complexity of ESG-related challenges. Integrated functions allow internal auditors to proactively identify and address ESG risks before they escalate. Internal auditors must not only provide assurance on the efficacy and integrity of existing ESG processes in order to identify areas of strength and weakness, but also design and improve comprehensive ESG frameworks that are suited to the organisation's unique needs and objectives (IIA, 2021). This entails working with key stakeholders to define specific ESG goals, targets, and performance metrics. Internal auditors can use their knowledge of risk management, governance, and sustainability to create complete frameworks that address environmental, social, and governance issues successfully.

Effective internal auditing also requires providing strategic consulting services on how to navigate complex ESG challenges. This involves advising board members and senior management on emerging ESG trends, regulatory developments and stakeholder expectations (Feeney & Aiken, 2022; Rakipi & D'Onza, 2023). Internal auditors can provide information on industry standards, cutting-edge ESG management techniques, and best practices. Internal audit teams should utilise their in-depth understanding and knowledge of the industry to contribute to well-informed decision-making and planning for strategic integration of ESG factors.

11.3 Discussion of Implications

The adoption of the ABC model in internal audit presents significant implications for businesses, governance structures, and regulatory frameworks. This section discusses the practical, theoretical, and strategic implications of the ABC Model by analysing its impact across different industries, particularly in ESG-related risks, IT governance, and supply chain governance.

11.3.1 Practical Implications

The ABC model reshapes internal auditing, extending beyond traditional assurance to active participation in business growth and strategy. This shift requires businesses to reconsider how internal auditors contribute to risk management, regulatory compliance, and strategic decision-making. The implications of this shift are illustrated through applications in different industries.

11.3.1.1 Example 1: Utilising the ABC Model in ESG-Related Risks

Let's consider an example involving an oil and gas company implementing an ESG program focused on reducing greenhouse gas emissions. In the oil and gas industry, internal auditors play a critical role in ensuring compliance with environmental regulations and addressing ESG concerns. For instance, Malaysia's Petronas launched its Net Zero Carbon Emissions by 2050 agenda, which includes initiatives such as methane emissions reduction, carbon capture and storage (CCS), and renewable energy investments (Nizam, 2022). Similarly, Thailand's PTT Group has committed to achieving Net Zero GHG Emissions by 2050, focusing on renewable energy expansion, carbon offset programs, and innovation in sustainable technologies (PTT Group, 2023). Additionally, multinational corporations, such as Shell and ExxonMobil, which operate extensively across Asia, have implemented regional ESG strategies, including carbon reduction roadmaps, flaring reduction initiatives, and biodiversity conservation projects in their Asian operations (ExxonMobil, 2023; Shell, 2023). These examples illustrate how both regional and global players integrate ESG programs into their operations across Asia in response to regulatory pressures and stakeholder expectations for sustainable practices.

(1) *Assurance*: Internal auditors conduct audits to assess the company's emissions monitoring and reporting processes. They verify the accuracy and completeness of data collection methods, ensuring compliance with regulatory requirements such as those outlined in the Kyoto Protocol or Paris Agreement. By providing assurance on emissions data accuracy and transparency, internal auditors help the company build trust with stakeholders and mitigate reputational risks associated with environmental non-compliance.
(2) *Building*: Internal auditors collaborate with environmental engineers and sustainability experts to develop strategies for reducing greenhouse gas emissions across the company's operations. They participate in the design and implementation of initiatives such as methane leak detection and repair programs, flaring reduction projects, and investments in renewable energy sources. By actively contributing to the development of emission reduction strategies, internal auditors help the company achieve its sustainability goals while also enhancing operational efficiency and cost-effectiveness.
(3) *Consulting*: Internal auditors provide strategic advice to senior management on emerging trends and best practices in emissions reduction within the oil and gas industry. They offer insights into innovative technologies and industry benchmarks for reducing carbon intensity and improving environmental performance. By advising on the adoption of sustainable practices and investments in emission reduction initiatives, internal auditors help the company navigate the transition to a low-carbon economy and position itself as a responsible corporate citizen.

In this example, internal auditors in the oil and gas industry leverage the ABC model to provide *assurance, build* robust emission reduction strategies, and offer strategic *consulting* services, ultimately contributing to the company's ESG objectives, long-term sustainability, and future-proofing its business. By integrating the ABC model into their audit practices, the internal auditors ensure compliance with environmental regulations, actively participate in developing sustainability initiatives, and advise senior management on emerging trends and best practices. This holistic approach not only addresses current ESG challenges but also prepares the company to navigate future uncertainties, ensuring its resilience and continued success in the ever-evolving business landscape.

11.3.1.2 Example 2: Utilising the ABC Model in IT Governance

The integration of the ABC model in continuous auditing within an IT environment offers substantial benefits, particularly when leveraging technological solutions. Internal audit, with its familiarity with company systems and extensive data access, gains a competitive advantage over external audit by employing Big Data Analytics (BDA) (Betti & Sarens, 2021; Betti et al., 2023). Furthermore, internal audit is well-suited for continuous auditing practices, such as process mining, which enables auditors to extract valuable insights from large datasets. By analysing extensive data, auditors can visualise processes, evaluate performance, and identify inefficiencies across various IT systems, enhancing operational understanding and facilitating the detection of discrepancies (Lenz, 2017). This integration empowers auditors to conduct thorough and efficient audits, ultimately improving risk management and governance within the IT environment.

For example, the Singapore-based DBS Bank has implemented robust IT governance frameworks that integrate data analytics and AI-driven monitoring tools to enhance risk management and cybersecurity resilience (DBS Group, 2023). Similarly, Malaysia's Maybank has adopted digital risk governance frameworks to oversee its cloud computing strategies and ensure regulatory compliance across IT operations (Maybank, 2023). In addition, global companies, such as IBM and Microsoft, which maintain significant operations and client bases across Asia, support IT governance in the region through the deployment of AI-powered risk assessment tools, process mining solutions, and continuous auditing systems tailored for Asian markets (IBM, 2023; Microsoft, 2023). These examples reflect the increasing adoption of advanced IT governance practices across Asia to safeguard digital assets and to support business continuity.

Let's consider an example of a multinational financial institution that implements the ABC model to enhance its internal audit function, particularly in continuous auditing within its IT systems.

(1) *Utilising Big Data Analytics (BDA)*: The internal audit team, leveraging its access to vast amounts of transactional data and IT systems, employs BDA techniques to analyse customer transactions, detect anomalies, and identify potential instances of fraud or financial irregularities. By harnessing BDA tools, such as predictive analytics and machine learning algorithms, auditors can sift through immense datasets efficiently and effectively, gaining insights that may not be possible through traditional audit methods alone.
(2) *Continuous Auditing with Process Mining*: Through the ABC model, internal auditors utilise process mining techniques to analyse the end-to-end flow of financial transactions within the institution's IT systems. By visualising and mapping out the various processes, auditors can identify bottlenecks, inefficiencies, and control weaknesses in the system. For example, they may discover delays in transaction processing, deviations from established controls, or unauthorised access to sensitive data. This enables auditors to promptly address issues, strengthen controls, and improve operational efficiency.
(3) *Enhanced Risk Management and Governance*: By integrating the ABC model with continuous auditing practices within the IT environment, the financial institution can significantly enhance its risk management and governance processes. Auditors can proactively identify and mitigate risks, ensuring compliance with regulatory requirements and safeguarding the integrity of financial operations. Additionally, by leveraging technology-driven audit approaches, the financial institution can streamline audit processes, reduce manual effort, and enhance audit coverage, ultimately contributing to improved overall governance and risk management.

In this second example, the integrated ABC model empowers internal auditors to effectively leverage technological solutions for continuous auditing within the IT environment, contributing to business future-proofing. By harnessing the capabilities of the ABC model, auditors can proactively identify risks, strengthen governance frameworks, and enhance operational efficiency within the financial institution. This strategic approach not only ensures compliance with regulatory requirements but also enables the institution to adapt to evolving market conditions and emerging risks, ultimately safeguarding its long-term viability and resilience.

11.3.1.3 Example 3: Utilising the ABC Model in Supply Chain Governance

Supply chain governance has grown in importance as corporations are held accountable for their ESG practices and the ESG standards maintained by their suppliers and partners. Internal auditors can play an important role in improving supply chain governance by assuring compliance, identifying weaknesses, and promoting sustainable practices throughout the supply chain.

For example, Japan's Toyota has implemented Supplier Sustainability Guidelines to promote ethical sourcing, environmental responsibility, and human rights compliance across its global supply chains, including operations in Asia (Toyota, 2023). Malaysia's Sime Darby Plantation introduced a Responsible Agriculture Charter, focusing on traceability, fair labour practices, and zero-deforestation commitments within its palm oil supply chain (Sime Darby Plantation, 2023). Additionally, global companies, such as Unilever, with a strong operational presence in Asia, have adopted the Responsible Sourcing Policy to ensure that their suppliers meet strict ESG standards, covering labor rights, environmental protection, and anti-corruption measures (Unilever, 2023). These initiatives exemplify how both Asian companies and MNCs embed ESG principles in supply chain governance across the region.

Consider a retail company that is rolling out an ESG program aimed at promoting sustainable supply chain practices. In the retail sector, internal auditors are crucial in ensuring that the company's operations and its suppliers comply with ESG standards across the entire supply chain.

(1) *Assurance*: Internal auditors in a retail organisation can implement supplier audits to ensure compliance with ESG standards which includes adherence of suppliers to fair labour practices, environmental regulations, and ethical sourcing policies. For example, internal auditors may review supply chain data to confirm that apparel suppliers meet sustainability and labour laws in the countries where products are manufactured. By providing this assurance, internal audit helps the retail organisation maintain stakeholder trust and avoid reputational risks associated with non-compliance.
(2) *Build*: In building a resilient and transparent supply chain, internal auditors collaborate with procurement and supply chain teams to create comprehensive ESG frameworks for supplier selection and performance assessment. This may

involve establishing sustainability metrics specific to sourcing practices, such as minimising carbon footprint in product transportation or ensuring waste reduction in packaging. Internal auditors may also help design due diligence processes for new suppliers, ensuring that ESG criteria are embedded into vendor agreements and procurement standards. This proactive approach aligns the retail organisation's supply chain operations with its ESG objectives and values.

(3) *Consulting*: Internal auditors, with their expertise in risk assessment and compliance, are the best parties to advise senior management on best practices with regards to responsible sourcing and fair trade. By using their knowledge of regulations and market trends, internal auditors can suggest strategies like using reusable packaging, sourcing eco-friendly materials, or adopting technologies to track product origins. These insights help the retail company keep up with changing consumer expectations and new sustainability rules, ensuring the company stays competitive and responsible.

11.3.2 Theoretical Implications

The ABC Model challenges conventional theories in internal audit by expanding its role beyond assurance and consulting to include building, positioning internal auditors as value creators rather than mere oversight agents. This shift has several theoretical implications:

(1) Extending Agency Theory—Traditional internal auditing is based on agency theory, ensuring compliance and risk mitigation for shareholders. However, the ABC Model shifts internal audit's role to a strategic function, balancing the interests of multiple stakeholders beyond shareholders, including regulators, environmental agencies, and communities.

The ABC Model expands internal audit's role beyond passive oversight to active engagement in business resilience and strategic decision-making. The "Build" function enables auditors to collaborate with management to enhance governance structures, ESG compliance, and risk management frameworks. This challenges the traditional agent-principal perspective by positioning auditors as partners in value creation rather than just compliance enforcers. For example, in ESG-related risk management, auditors do not just verify compliance but help build sustainability initiatives, ensuring long-term shareholder value.

(2) Enhancing Resource Dependence Theory—Resource Dependence Theory suggests that organisations rely on external and internal resources to manage risks and operational efficiency. Internal audit was traditionally seen as a support function rather than a strategic resource.

The ABC Model positions internal audit as a critical resource that contributes to an organisation's strategic growth and resilience. The "Build" function makes auditors active participants in developing governance frameworks, digital transformation, and ESG strategies. Internal auditors become key organisational assets who enhance business continuity, cybersecurity risk management, and innovation. For instance, in IT governance, auditors do not just assess risks but also collaborate with IT teams to build cybersecurity and data protection strategies.

(3) Reframing Stakeholder Theory—Stakeholder Theory argues that businesses must consider the interests of all stakeholders (not just shareholders) in decision-making. Internal audit traditionally focused on financial integrity and regulatory compliance, primarily serving investors and regulators.

The ABC Model aligns internal audit with broader stakeholder needs, including employees, customers, suppliers, and the environment. The "Build" function enables internal auditors to support sustainability, ethical supply chains, and digital trust. Internal auditors become governance enablers who bridge the gap between corporate objectives and stakeholder expectations. For example, in supply chain governance, auditors ensure ethical sourcing, labor rights compliance, and sustainability—not just regulatory adherence.

By integrating these theoretical frameworks, the ABC Model modernises internal auditing, positioning it as a key organisational enabler rather than just a compliance mechanism.

11.4 Future Directions

11.4.1 Strategic Directions

The strategic directions of the ABC Model highlight how businesses, regulators, and policymakers must adapt to the evolving role of internal audit in governance, risk management, and sustainability. This sub-section explains how the expanded role of internal audit impacts corporate strategy, regulatory frameworks, and technological advancements.

11.4.1.1 Regulatory Adaptation

As internal auditors expand beyond compliance to strategic governance, regulators must update policies and governance frameworks to align with the evolving responsibilities of internal audit.

ESG Compliance: With increasing global emphasis on sustainability and ESG reporting, businesses need internal auditors to provide independent ESG assurance. For example, Governments and regulators (e.g., the Securities Commission, Bursa Malaysia, and EU Corporate Sustainability Reporting Directive) are implementing stricter ESG disclosure regulations. In Asia, regulators such as Singapore Exchange (SGX) require listed companies to publish sustainability reports aligned with TCFD recommendations, where internal audit contributes by verifying the accuracy and completeness of disclosures. Hong Kong's Stock Exchange (HKEX) has also strengthened ESG reporting standards, prompting internal auditors to assess ESG risk integration into corporate strategies. In Japan, the revised Corporate Governance Code encourages boards to consider sustainability issues, thereby expanding the role of internal auditors in providing assurance on ESG-related risks and performance. The ABC Model helps companies navigate compliance by integrating ESG audits into risk management frameworks.

Cybersecurity & IT Governance: Regulatory bodies are tightening cybersecurity laws to mitigate risks associated with data breaches and financial fraud. For example, the implementation of GDPR (General Data Protection Regulation) in the EU and Malaysia's Personal Data Protection Act (PDPA) requires internal auditors to play a proactive role in IT risk assessments. Asian countries like India have enacted the Digital Personal Data Protection Act (DPDPA) 2023, where internal audit is essential for assessing data lifecycle controls, breach readiness, and compliance with consent protocols. In China, the Cybersecurity Law and Data Security Law (DSL) impose strict governance over personal data and cross-border data flows, where internal auditors must evaluate controls over data classification, protection, and usage.

11.4.1.2 Technology-Driven Internal Audit

With AI, big data analytics, and blockchain transforming business risk management, internal auditors must integrate advanced technologies to enhance real-time auditing and fraud detection.

AI & Big Data Analytics: Internal auditors are leveraging AI-powered analytics for continuous auditing and anomaly detection. For example, financial institutions use big data analytics and machine learning to monitor financial transactions in real-time and prevent fraud. ABC Model Implication: The "Build" function requires auditors to collaborate with IT and data science teams to enhance predictive risk assessment.

Blockchain & Smart Contracts: Blockchain technology enables tamper-proof record-keeping, improving transparency in financial reporting and supply chain governance. For example, retail and manufacturing companies use blockchain for real-time supply chain tracking, ensuring ethical sourcing compliance. ABC Model Implication: Auditors must consult on blockchain implementation strategies and provide assurance on digital assets and transactions.

11.4.1.3 Redefining the Role of Internal Auditors

The ABC Model transforms internal auditors into strategic advisors, requiring new skill sets in sustainability, digital risk management, and corporate strategy. Businesses must redesign training programs to equip auditors with multi-disciplinary expertise.

ESG & Sustainability Auditing: Internal auditors must assess corporate ESG initiatives and provide recommendations on climate risks, carbon reduction strategies, and ethical business practices. For example, retail auditors must verify supply chain sustainability compliance with global fair trade and environmental standards.

Cybersecurity & IT Risk Expertise: Businesses require internal auditors to assess IT controls, cybersecurity risks, and regulatory compliance for digital operations. For example, financial auditors must evaluate cloud security risks and advise on cyber resilience strategies.

11.4.2 Research Directions

Looking forward, the future direction of the ABC model in internal audit should involve even greater integration with organisational strategy and a more proactive stance in identifying and addressing risks and opportunities (Christ et al., 2021; Handoyo, 2024; Kotb et al., 2020). There are several potential areas for future exploration and refinement of the ABC model within the internal audit role and function. First, the role of advanced technologies such as artificial intelligence, data analytics, and blockchain in supporting the ABC model should be actively explored to enhance the efficiency and effectiveness of internal audit activities in all 3 dimensions of the model. Internal auditors should be encouraged to investigate how technology can support assurance activities, contribute to building robust frameworks and facilitate strategic consulting (Betti & Sarens, 2021; Betti et al., 2023; Christ et al., 2021).

Second, firms should invest in further research to refine the ABC model, explore specific industries, organisational sizes, and contexts to better tailor its application (Bonrath et al., 2022; DeSimone et al., 2021; Shuwaili et al., 2023). For example, firms can collaborate with professional bodies (e.g. IIA) and regulatory authorities to develop industry-specific best practices and case studies to guide internal auditors in diverse environments (Eulerich & Lenz, 2020; Feeney & Aiken, 2022).

Third, firms can develop comprehensive training programs and initiate professional development opportunities to equip internal auditors with the necessary skills and knowledge for success in each dimension of the ABC model (Amoako et al., 2023; DeSimone et al., 2021). Finally, a culture of continuous learning and adaptation that embraces the expanded role of internal audit should be fostered within the internal audit teams (Betti et al., 2023; Christ et al., 2021; Handoyo, 2024). By doing so, internal auditors will be better equipped to navigate the evolving business landscape and address emerging risks effectively. In conclusion, the ABC model opens exciting possibilities for the future of internal audit and facilitates future-proofing business.

11.5 Conclusion

In summary, the ABC model is a useful approach for internal auditors to cope with the changing business landscape. It is an innovative framework to redefine and expand the role of the internal audit function (Lenz & Hoos, 2023), from a primarily assurance-focused function to one that actively contributes to building and enhancing organisational capabilities. By ensuring the integrity and compliance of financial reporting ('Assure'), actively participating in building robust governance structures and risk management frameworks that safeguard assets ('Build'), and providing strategic advice to improve decision-making processes ('Consult'), the ABC model helps internal auditors to position themselves as invaluable partners in governance, risk management and organisational development (Betti et al., 2023; Lenz & Chesshire, 2023; Lenz & Hoos, 2023).

This chapter contributes by explaining how internal auditors from three different industries (oil and gas; financial; and retail) utilise the ABC model in their ESG-related risk assessments, reporting requirements, and governance in IT and supply chain. In so doing, two key strengths of the ABC model is substantiated. First, this model acknowledges the evolving nature of the business landscape and the need for internal auditors to go beyond their traditional boundaries (Handoyo, 2024; Lenz, 2017; Lenz & Chesshire, 2023). Second, this model recognises the multi-faceted nature of the internal audit function, emphasising not only risk mitigation and control assessment but also active involvement in fostering innovation, resilience, and strategic growth (Christ et al., 2021; Rakipi & D'Onza, 2023).

While the ABC model holds promise for a more proactive and value-driven internal audit role and function, it is not without limitations. One key constraint is the potential for role ambiguity, limited resources and skill gaps within internal audit teams (Alqudah et al., 2023; Amoako et al., 2023). Striking the right balance between traditional assurance activities and the expanded roles of building and consulting requires careful resource allocation and continuous skill development. For instance, internal auditors may find it challenging to balance their objectivity and independence with ESG expectations and stakeholder demands (Rakipi & D'Onza, 2023). Hence, successfully navigating the expanded roles of internal audit requires a nuanced understanding of the balance between assurance, building, and consulting (Lenz & Chesshire, 2023; Lenz & Hoos, 2023; Lenz & Jeppesen, 2022). Next, resistance to change, lack of leadership or rigid organisational structures may impede the successful adoption of the ABC model. Therefore, it is imperative to recognise that the applicability of the model may vary across different industries, organisational cultures and leadership support (Alqudah et al., 2023; Handoyo, 2024; Shuwaili et al., 2023).

Despite these challenges, the ABC model opens exciting possibilities for the future of internal audit, positioning auditors as key enablers of strategic decision-making, risk management, and governance. By embracing technological advancements, ESG integration, and continuous learning, internal auditors can future-proof businesses and create long-term value in an increasingly complex and dynamic environment.

Highlights

- Adopt the Assure-Build-Consult (ABC) Model to move beyond compliance, enabling Asian audit functions to anticipate challenges and co-create value with line management.
- Integrate environmental, social, and governance criteria into every audit cycle to meet tightening global standards and rising stakeholder demands, securing long-term resilience in Asian markets.
- Harness AI, data analytics, and blockchain for real-time, predictive assurance—responding to Asia's accelerated digitalization and the urgency for rapid, data-driven decision-making.
- Design audit methodologies that reflect Asia's spectrum of industries and market maturities—from emerging economies to developed hubs—ensuring relevance and practical impact.
- Move beyond threat detection by embedding scenario planning and strategic advisory into audit plans, empowering enterprises to navigate emerging risks and dynamic regulatory landscapes.
- Develop multidisciplinary audit teams with expertise in digital tools, critical thinking, and business strategy, equipping Asia's internal audit professionals to adapt swiftly to regulatory and technological shifts.

References

Alqudah, H., Amran, N. A., Hassan, H., Lutfi, A., Alessa, N., Alrawad, M., & Almaiah, M. A. (2023). Examining the critical factors of internal audit effectiveness from internal auditors' perspective: Moderating role of extrinsic rewards. *Heliyon, 9*(10), Article e20497. https://doi.org/10.1016/j.heliyon.2023.e20497

Amoako, G. K., Bawuah, J., Asafo-Adjei, E., & Ayimbire, C. (2023). Internal audit functions and sustainability audits: Insights from manufacturing firms. *Cogent Business & Management, 10*(1), 2192313. https://doi.org/10.1080/23311975.2023.2192313

Betti, N., DeSimone, S., Gray, J., & Poncin, I. (2023). The impacts of the use of data analytics and the performance of consulting activities on perceived internal audit quality. *Journal of Accounting & Organizational Change*, ahead-of-print(ahead-of-print). https://doi.org/10.1108/JAOC-08-2022-0125

Betti, N., & Sarens, G. (2021). Understanding the internal audit function in a digitalised business environment. *Journal of Accounting & Organizational Change, 17*(2), 197–216. https://doi.org/10.1108/JAOC-11-2019-0114

Bonrath, A., Eulerich, M., & Lopez-Kasper, V. (2022). Internal auditor's role in ESG disclosure and assurance: An analysis of practical insights. *Corporate Ownership & Control*, *20*(1), 78–86. https://doi.org/10.22495/cocv20i1art7

Christ, M. H., Eulerich, M., Krane, R., & Wood, D. A. (2021). New frontiers for internal audit research. *Accounting Perspectives, 20*(4), 449–475. https://doi.org/10.1111/1911-3838.12272

DBS Group. (2023). *DBS Annual Report 2023: Strengthening Digital and Cyber Resilience.* https://www.dbs.com/investor/annual-report.html

DeSimone, S., D'Onza, G., & Sarens, G. (2021). Correlates of internal audit function involvement in sustainability audits. *Journal of Management and Governance, 25*(2), 561–591. https://doi.org/10.1007/s10997-020-09511-3

Eulerich, M., & Lenz, R. (2020). *Defining, measuring, and communicating the value of internal audit: Best practices for the profession.* Internal Audit Foundation.

ExxonMobil. (2023). *ExxonMobil Sustainability Report 2023.* https://corporate.exxonmobil.com/sustainability-and-reports/sustainability

Feeney, C., & Aiken, J. (2022). *Step-by-step guide to building your ESG Program: Resources, best practices, and key considerations.* AuditBoard Inc. https://www.auditboard.com/resources/ebook/step-by-step-guide-to-building-your-esg-program/

Gramling, A. A., Maletta, M. J., Schneider, A., & Church, B. K. (2004). The role of the internal audit function in corporate governance: A synthesis of the extant internal auditing literature and directions for future research. *Journal of Accounting Literature, 23*(1), 194–244.

Gupta, P. P., & Ray, M. R. (1992). The changing roles of the internal auditor. *Managerial Auditing Journal, 7*(1), 3–8. https://doi.org/10.1108/EUM0000000001770

Handoyo, S. (2024). Mapping the landscape of internal auditing effectiveness study: A bibliometric approach. *Cogent Business & Management, 11*(1), 2289200. https://doi.org/10.1080/23311975.2023.2289200

IBM. (2023). *IBM IT Governance and Risk Solutions.* https://www.ibm.com/security/it-governance

IIA. (2021). Internal audit's role in ESG reporting: Independent assurance is critical to effective sustainability reporting. In *White paper.* The Institute of Internal Auditors.

IIPF. (2017). *International professional practices framework.* The Institute of Internal Auditors.

Kotb, A., Elbardan, H., & Halabi, H. (2020). Mapping of internal audit research: A post-Enron structured literature review. *Accounting, Auditing & Accountability Journal, 33*(8), 1969–1996. https://doi.org/10.1108/AAAJ-07-2018-3581

Lenz, R. (2017). Time is ripe to revolutionize the audit. *EDPACS, 56*(4), 19–22. https://doi.org/10.1080/07366981.2017.1380479

Lenz, R., & Chesshire, J. (2023). Rethinking internal audit: Governance needs gardening. *EDPACS, 68*(3), 7–15. https://doi.org/10.1080/07366981.2023.2255432

Lenz, R., & Hoos, F. (2023). The future role of the internal audit function: Assure. Build. Consult. *EDPACS, 67*(3), 39–52. https://doi.org/10.1080/07366981.2023.2165361

Lenz, R., & Jeppesen, K. K. (2022). The future of internal auditing: Gardener of governance. *EDPACS, 66*(5), 1–21. https://doi.org/10.1080/07366981.2022.2036314

Lenz, R., & Sarens, G. (2012). Reflections on the internal auditing profession: What might have gone wrong? *Managerial Auditing Journal, 27*(6), 532–549. https://doi.org/10.1108/02686901211236382

Maybank. (2023). *Maybank sustainability and digital governance report 2023.* https://www.maybank.com/iwov-resources/documents/pdf/annual-report/2023/Maybank-Sustainability-Report-2023.pdf

Microsoft. (2023). *Microsoft risk management and compliance solutions.* https://www.microsoft.com/en-us/security/blog/

Nizam, F. (2022). *Petronas committed to net zero carbon emissions by 2050.* https://www.nst.com.my/news/nation/2022/12/861164/petronas-committed-net-zero-carbon-emissions-2050

Prawitt, D. F., Smith, J. L., & Wood, D. A. (2009). Internal audit quality and earnings management. *The Accounting Review, 84*(4), 1255–1280. https://doi.org/10.2308/accr.2009.84.4.1255

PTT Group. (2023). *PTT Group Sustainability Report 2023.* https://sustainability.pttgcgroup.com/storage/document/integrated-sustainability-reports/2023/pttgc-isr2023.pdf

Rakipi, R., & D'Onza, G. (2023). The involvement of internal audit in environmental, social, and governance practices and risks: Stakeholders' salience and insights from audit committees and chief executive officers. *International Journal of Auditing*, 1–14. https://doi.org/10.1111/ijau.12341

Ridley, J., D'Silva, K., & Szombathelyi, M. (2011). Sustainability assurance and internal auditing in emerging markets. *Corporate Governance: The International Journal of Business in Society, 11*(4), 475–488. https://doi.org/10.1108/14720701111159299

Roussy, M. (2013). Internal auditors' roles: From watchdogs to helpers and protectors of the top manager. *Critical Perspectives on Accounting, 24*(7), 550–571. https://doi.org/10.1016/j.cpa.2013.08.004

Roussy, M., & Perron, A. (2018). New perspectives in internal audit research: A structured literature review. *Accounting Perspectives, 17*(3), 345–385. https://doi.org/10.1111/1911-3838.12180

Sarens, G., De Beelde, I., & Everaert, P. (2009). Internal audit: A comfort provider to the audit committee. *The British Accounting Review, 41*(2), 90–106. https://doi.org/10.1016/j.bar.2009.02.002

Shell. (2023). *Shell Sustainability Report 2023*. https://www.shell.com/sustainability.html

Shuwaili, A. M. J., Hesarzadeh, R., & Velashani, M. A. B. (2023). Designing an internal audit effectiveness model for public sector: Qualitative and quantitative evidence from a developing country. *Journal of Facilities Management*, ahead-of-print(ahead-of-print). https://doi.org/10.1108/JFM-07-2022-0077

Sime Darby Plantation. (2023). *Responsible agriculture charter and sustainability commitments*. https://www.worldclimatefoundation.org/post/sime-darby-plantation-s-commitment-to-forest-and-biodiversity-conservation#:~:text=Our%20Responsible%20Agriculture%20Charter%20(RAC,development%2C%20poaching%2C%20and%20hunting

Soh, D. S. B., & Martinov-Bennie, N. (2011). The internal audit function. *Managerial Auditing Journal, 26*(7), 605–622. https://doi.org/10.1108/02686901111151332

Soh, D. S. B., & Martinov-Bennie, N. (2015). Internal auditors' perceptions of their role in environmental, social and governance assurance and consulting. *Managerial Auditing Journal, 30*(1), 80–111. https://doi.org/10.1108/MAJ-08-2014-1075

Stewart, J., & Subramaniam, N. (2010). Internal audit independence and objectivity: Emerging research opportunities. *Managerial Auditing Journal, 25*(4), 328–360. https://doi.org/10.1108/02686901011034162

Toyota. (2023). *Toyota supplier sustainability guidelines*. https://www.toyota-industries.com/sustainability/item/Supplier%20Sustainability%20Guidelines_E.pdf

Unilever. (2023). *Responsible sourcing policy*. https://www.unilever.com/sustainable-living/responsible-sourcing/

Vinnari, E., & Skærbæk, P. (2014). The uncertainties of risk management. *Accounting, Auditing & Accountability Journal, 27*(3), 489–526. https://doi.org/10.1108/AAAJ-09-2012-1106

Suhaily Shahimi is a Senior Lecturer at the Department of Accounting, Faculty of Business and Economics, Universiti Malaya (Malaysia), holds a Ph.D. in Internal Audit and Corporate Governance. She brings extensive expertise in education management with a strong focus on Internal Audit, Risk Management, and Governance. Dr. Suhaily teaches advanced auditing, internal auditing, integrated case studies, and research methodology for MBA and Master of Accounting programs. Her research spans auditing, risk management, governance, ESG, data analytics, and Artificial Intelligence. Recognised for her contributions, she earned the Best Presenter and Best Paper Awards at the 2021 International Conference on Accounting Research and Education.

Mohd Haniff Zainuldin is a Senior Lecturer at the Department of Accounting, Universiti Malaya (Malaysia). He holds a Ph.D. in Accounting from the University of South Australia. Mohd Haniff has been an academic for more than 14 years in various private and public higher education institutions. He has published widely in top ranked accounting and finance journals such as Corporate Social Responsibility and Environmental Management, International Journal of Bank Marketing and Pacific Basin Finance Journal. His research interests include capital market-based research, Environmental, Social and Governance (ESG) as well as Islamic finance principles.

Elaine Yen Nee Oon is a Senior Lecturer at the Accounting Department, Faculty of Business & Economics, Universiti Malaya (Malaysia), with an M.Phil and D.Phil from Cambridge University, UK. A former World Bank scholar, she is a Chartered Accountant and Certified Financial Planner. She published in Management Decision, Managerial and Decision Economics, Frontiers in Psychology, Cogent Economics & Finance and others. Elaine received the Best Reviewer Award 2020 and 2021 (International Management Division, Academy of Management); the Best Paper Award IMAC 2021 and Emerald-IMAC 2024 Most Provocative Research Award. She currently serves on the Editorial Board of Asian Journal of Business and Accounting.

Index

M. S. Balaji et al. (eds.), *Future of Business in Asia*,
https://doi.org/10.1007/978-3-031-95940-0